Studies in Logic
Volume 83

Reasoning:
Games, Cognition, Logic

Volume 73
Measuring Inconsistency in Information
John Grant and Maria Vanina Martinez, eds.

Volume 74
Dictionary of Argumentation. An Introduction to Argumentation Studies
Christian Plantin. With a Foreword by J. Anthony Blair

Volume 75
Theory of Effective Propositional Paraconsistent Logics
Arnon Avron, Ofer Arieli and Anna Zamansky

Volume 76
Argumentation and Inference. Proceedings of the 2nd European Conference on Argumentation. Volume I
Steve Oswald and Didier Maillat, eds.

Volume 77
Argumentation and Inference. Proceedings of the 2nd European Conference on Argumentation. Volume II
Steve Oswald and Didier Maillat, eds.

Volume 68
Logic and Philosophy of Logic. Recent Trends in Latin America and Spain
Max A. Freund, Max Fernández de Castro and Marco Ruffino, eds.

Volume 79
Games Iteration Numbers. A Philosophical Introduction to Computability Theory
Luca M. Possati

Volume 80
Logics of Proofs and Justifications
Roman Kuznets and Thomas Studer

Volume 81
Factual and Plausible Reasoning
David Billington

Volume 82
Formal Logic: Classical Problems and Proofs
Luis M. Augusto

Volume 83
Reasoning: Games, Cognition, Logic
Mariusz Urbański, Tomasz Skura and Paweł Łupkowski, eds.

Studies in Logic Series Editor
Dov Gabbay   dov.gabbay@kcl.ac.uk

Reasoning:
Games, Cognition, Logic

edited by
Mariusz Urbański
Tomasz Skura
Paweł Łupkowski

© Individual authors and College Publications, 2020
All rights reserved.

ISBN 978-1-84890-325-8

College Publications
Scientific Director: Dov Gabbay
Managing Director: Jane Spurr

http://www.collegepublications.co.uk

All rights reserved. No part of this publication may be reproduced, stored in a retrieval system or transmitted in any form, or by any means, electronic, mechanical, photocopying, recording or otherwise without prior permission, in writing, from the publisher.

Contents

Preface vii

Contributors ix

On two simple models for one simple game: 'Guess Who?', Inferential Erotetic Logic, and situational semantics 1
Mariusz Urbański and Joanna Grzelak

Natural Deduction Method for Solving CL-based Puzzles 19
Agata Tomczyk

The Measurement of Factive Deductivity: a Psychological and Cerebral Review 33
Paula Álvarez-Merino and Carmen Requena and Francisco Salto

Cognitive Principles and Individual Differences in Human Syllogistic Reasoning 55
Emmanuelle-Anna Dietz Saldanha and Richard Mörbitz

Acceptable propositional normal logic programs checking procedure implementation 77
Aleksandra Czyż, Kinga Ordecka, Andrzej Gajda

Propositional logic with probability operators (based on general ideas of weak modal calculus) 87
Tomasz Witczak

Let Me Ask You an Easier Question—Modifying and Rephrasing Questions in Information Seeking Dialogues 99
Paweł Łupkowski

Interleaved Argumentation and Explanation in Dialog 115
Adrian Groza

A Formal Ontology for Conception Representation in Terminological Systems 137
Farshad Badie

What is refutation? 157
Gabriele Pulcini and Tomasz Skura

Refutation systems in the finite 169
Valentin Goranko and Tomasz Skura

Towards a Uniform Account of Proofs and Refutations 181
Andrzej Wiśniewski

On Sequent-Type Rejection Calculi for Many-Valued Logics 189
Mihail Bogojeski and Hans Tompits

Preface

This volume contains papers presented at the Poznań Reasoning Week (PRW; https://poznanreasoningweek.wordpress.com/) multi-conference held in Poznań in September 11–15, 2018. PRW aims at bringing together experts, whose research offers a broad range of perspectives on systematic analyses of reasoning processes and their formal modelling. The 2018 edition consisted of three conferences, which addressed the following topics: (i) games in reasoning research, (ii) the interplay of logic and cognition, and (iii) refutation systems. The papers collected in this volume address all these topics.

Starting with games in reasoning research, the volume opens with the paper entitled *On two simple models for one simple game: 'Guess Who?', Inferential Erotetic Logic, and situational semantics* by Mariusz Urbański and Joanna Grzelak. Here we find a model for playing 'Guess Who?' game, developed within the framework of Inferential Erotetic Logic, subsequently refined in terms of situational semantics. The paper introduces the concept of discriminatory power of a polar question. Next paper by Agata Tomczyk: *Natural Deduction Method for Solving CL-based Puzzles*, presents a formal way of solving Smullyan's Combinatory logic-based puzzles by way of Curry-Howard isomorphism. The approach is formalized in natural deduction-style and uses unmodified and type-free combinators. Solving a puzzle is understood as creating a proof for a given formula, starting with a set of pre-assumed conditions.

Moving to the interplay between logic and cognition: the paper by Paula Álvarez-Merino, Carmen Requena, and Francisco Salto entitled *The Measurement of Factive Deductivity: a Psychological and Cerebral Review* presents a systematic review of psychometrical and neural tests, which were carried out in search for measures to empirically confirm or refute the Factive Deduction Hypothesis. Authors identify 27 psychometrical and 58 neural measures in their review. Emmanuelle-Anna Dietz Saldanha and Richard Mörbitz in *Cognitive Principles and Individual Differences in Human Syllogistic Reasoning* address the problem of human syllogistic reasoning and its formal modelling. The authors claim that an adequate cognitive theory of such a reasoning should aim at modeling individual differences among reasoners, if they are observed. They re-assess a novel approach based on the Weak Completion Semantics, Clustering by Principles, that addresses these differences. In *Acceptable propositional normal logic programs checking procedure implementation* Aleksandra Czyż, Kinga Ordecka and Andrzej Gajda propose solutions for three main issues that increase the computational cost of checking, if a given logic program is acceptable and present an implementation of the procedure written in Haskell programming language. Tomasz Witczak in *Propositional logic with probability operators (based on general ideas of weak modal calculus)* aims at establishing propositional calculus based on classical or intuitionistic core and equipped with probability operators.

In *Let Me Ask You an Easier Question—Modifying and Rephrasing Questions in Information Seeking Dialogues* by Paweł Łupkowski the author explores and analyses how questions are modified and rephrased in the information seeking dialogues in order to facilitate the answering process.

The next two papers apply description logic for constructing ontologies useful for analyses of human reasoning processes. Adrian Groza in his paper *Interleaved Argumentation and Explanation in Dialog* uses description logics in order to model argumentation and explanation. Description logics allow for defining the ontologies of the agents and also for distinguishing arguments from explanations. Farshad Badie in *A Formal Ontology for Conception Representation in Terminological Systems* applies description logics in order to offer a formal ontology for conception representation in terminological systems. This ontology specifies the conceptualisation of humans' conceptions as well as of the effects of their conceptions on the world.

The last four papers in this volume address issues related to refutation systems. In *What is refutation?* Gabriele Pulcini and Tomasz Skura present a logical analysis of the concept of refutation and illustrate some possible directions of research in the field of philosophical logic as well as in the methodology of propositional calculi. Valentin Goranko and Tomasz Skura in the paper *Refutation systems in the finite* present some refutation systems on finite semantic structures and establish some basic facts about them. In particular, the authors develop generic refutation systems for modal logics and for first-order theories that are semantically determined by single finite structures or by classes of finite structures, for arbitrary first-order languages. Andrzej Wiśniewski characterises a sequent calculus for holistically inconsistent sets of well formed formulas of Classical Propositional Logic in *Towards a Uniform Account of Proofs and Refutations*. A uniform method for constructing sequent-style rejection calculi for any given propositional finitely many-valued logic defined by means of a truth-functional semantics is proposed by Mihail Bogojeski and Hans Tompits in the paper entitled *On Sequent-Type Rejection Calculi for Many-Valued Logics*.

The editors of this volume would like to thank the authors for their contributions and the reviewers for their valuable feedback and insights on the papers. We would also like to thank the institutions which supported the PRW 2018 financially: National Science Centre (the grant "Modeling of abductive reasoning", DEC-2013/10/E/HS1/00172), Adam Mickiewicz University and University of Zielona Góra as well as for endorsing our initiative, Games Research Association of Poland, Polish Association for Logic and Philosophy of Science and Polish Society for Cognitive Science.

Mariusz Urbański, Tomasz Skura, Paweł Łupkowski

Contributors

FARSHAD BADIE, Center for Natural and Formal Languages, Aalborg University, Denmark, *badie@id.aau.dk*

MIHAIL BOGOJESKI, Institute of Software Engineering and Theoretical Computer Science, Technische Universität Berlin, Germany, *mihail.bogojeski@campus.tu-berlin.de*

ALEKSANDRA CZYŻ, Reasoning Research Group, Faculty of Psychology and Cognitive Science, Adam Mickiewicz University, Poznań, Poland, *aleksandra.m.czyz@gmail.com*

EMMANUELLE-ANNA DIETZ SALDANHA, International Center for Computational Logic, TU, Dresden, Germany, *emmanuelle.dietz@tu-dresden.de*

ANDRZEJ GAJDA, Department of Logic and Cognitive Science, Faculty of Psychology and Cognitive Science, Adam Mickiewicz University, Poznań, Poland, *andrzej.gajda@amu.edu.pl*

VALENTIN GORANKO, Stockholm University, Sweden, University of Johannesburg (visiting professorship), *valentin.goranko@philosophy.su.se*

ADRIAN GROZA, Department of Computer Science, Technical University of Cluj-Napoca, Romania, *Adrian.Groza@cs.utcluj.ro*

JOANNA GRZELAK, Reasoning Research Group, Faculty of Psychology and Cognitive Science, Adam Mickiewicz University, Poznań, Poland, *jomgrzelak@gmail.com*

PAWEŁ ŁUPKOWSKI, Department of Logic and Cognitive Science, Faculty of Psychology and Cognitive Science, Adam Mickiewicz University, Poznań, Poland, *Pawel.Lupkowski@amu.edu.pl*

PAULA ÁLVAREZ-MERINO, Universidad de León, Spain, *paulaalvarezmerino@gmail.com*

RICHARD MÖRBITZ, International Center for Computational Logic, TU, Dresden, Germany, *richard.moerbitz@tu-dresden.de*

KINGA ORDECKA, Reasoning Research Group, Faculty of Psychology and Cognitive Science, Adam Mickiewicz University, Poznań, Poland, *kingaordecka@gmail.com*

GABRIELE PULCINI, Institute for Logic, Language and Computation, University of Amsterdam, Holand, *g.pulcini@uva.nl*

CARMEN REQUENA, Universidad de León, Spain, *c.requena@unileon.es*

FRANCISCO SALTO, Universidad de León, Spain, *fsala@unileon.es*

TOMASZ SKURA, University of Zielona Góra, Poland, *T.Skura@ifil.uz.zgora.pl*

AGATA TOMCZYK, Department of Logic and Cognitive Science, Faculty of Psychology and Cognitive Science, Adam Mickiewicz University, Poznań, Poland, *a.tomczyk@protonmail.com*

HANS TOMPITS, Institute for Logic and Computation, Technische Universität Wien, Austria, *tompits@kr.tuwien.ac.at*

MARIUSZ URBAŃSKI, Department of Logic and Cognitive Science, Faculty of Psychology and Cognitive Science, Adam Mickiewicz University, Poznań, Poland, *mariusz.urbanski@amu.edu.pl*

ANDRZEJ WIŚNIEWSKI, Department of Logic and Cognitive Science, Faculty of Psychology and Cognitive Science, Adam Mickiewicz University, Poznań, Poland, *Andrzej.Wisniewski@amu.edu.pl*

TOMASZ WITCZAK, Institute of Mathematics, University of Silesia, Poland, *tm.witczak@gmail.com*

On two simple models for one simple game: 'Guess Who?', Inferential Erotetic Logic, and situational semantics

Mariusz Urbański and Joanna Grzelak

Abstract In this paper we introduce a model for playing 'Guess Who?' game developed within the framework of Inferential Erotetic Logic and then refine it in terms of situational semantics. We provide some comparison of the two models and shortly characterise well-known different strategies of playing the game in terms of the discriminatory power of polar questions.

Key words: 'Guess Who?', Inferential Erotetic Logic, logic of questions, polar question, discriminatory power, situational semantics, topic relevance

1 Introduction

In this paper we introduce two interrogative models for playing 'Guess Who?' game. The goal is to formally account for the dynamics of information processing, guided by consecutively asked questions and obtained answers. To this end, we introduce the concept of a discriminatory power of a polar question and of a history of a question in a 'Guess Who?' gameplay. The first model is elaborated in terms of Inferential Erotetic Logic (IEL; Wiśniewski 1995, 2013b), the second one in terms of a version of situational semantics (Wiśniewski, 2013a) (augmented with some additional concepts, in particular topic relevance), both based on Classical Propositional Calculus (CPC).

Mariusz Urbański
Department of Logic and Cognitive Science
Faculty of Psychology and Cognitive Science
Adam Mickiewicz University, Poznań, Poland
e-mail: mariusz.urbanski@amu.edu.pl

Joanna Grzelak
Reasoning Research Group
Faculty of Psychology and Cognitive Science
Adam Mickiewicz University, Poznań, Poland
e-mail: jomgrzelak@gmail.com

The IEL model offers adequate normative account on 'Guess Who?' gameplays, while the situational model enables more fine-grained analysis of semantic relations holding between questions.

The paper is organized as follows. We start with a short description of the game (Section 2). Then we introduce the concepts of discriminatory power of a polar question (Section 3) and of a history of a question in a 'Guess Who?' gameplay (Section 4). Further on, we describe our two models and offer some comparison between the two (Sections 5 and 6). We conclude with a short characteristics of well-known strategic concepts for playing 'Guess Who?' in terms of our framework. We also point at some possible directions for further research.

2 'Guess Who?'

'Guess Who?' is a two-player character guessing game created by Ora and Theo Coster, that was first manufactured by Milton Bradley, in 1979. The classic edition is currently being produced by Winning Moves (for more information see http://winning-moves.com/). The characters from this version of the game were depicted in the Figure 1 (further on we shall call them GW-characters; image downloaded from https://www.joe.co.uk/). There is a number of versions of the game available online[1].

Fig. 1: An exemplary set of 'Guess Who?' characters

---

[1] Some examples include: http://www.squiglysplayhouse.com, or http://www.agame.com.

'Guess Who?' may be played as a competitive game between two players. In such a setting each player choses a character and tries to identify the choice of the other player by questioning. The winner is the player who first identifies the opponent's choice correctly. In a non-competitive setting one player is the gamemaster, who choses the character, and the other player (or, possibly, players) tries to identify the character chosen. In both settings questions are often formulated as concerning the identity of the player themselves, and not a character (e. g. "Are you a woman?"). However, we shall employ also third-person wording (e. g. "Is the chosen character a woman?"). One important constraint is that only polar questions are allowed, that is, questions with two possible direct answers: 'yes' or 'no'. There is some discord as to what questions exactly are allowed in the game. While it is non-controversial that questions like "Do you have a beard?" can be used, it is not so obvious whether compound questions (e.g. "Do you have a beard or are you bald?") are legitimate as well. The same pertains to questions concerning, in a sense, meta-properties of the characters (e.g. "Are you in the top row of the picture?"). While important from the point of view of the game itself (its enjoyability, in particular), these legitimacy issues are of no concern from the point of view of our models. Also, they work equally well for competitive and non-competitive settings. Thus, for the sake of simplicity we shall be using examples of a non-competitive flavor, taken from a short pilot study carried out by the second author. One deviation from the rules we allowed for in this study was to ask about identity of the character chosen (e.g. "Are you Bernard?") at any stage of the game and not only to allow for such questions as the last ones in the gameplays.

In order to achieve the game's objective in the fewest number of steps the choice of questions to be asked is of crucial importance. 'Guess Who?' manuals often put this in a way for which the one from the Board Game Capital[2] is a good example:

> 'Guess Who?' is a classic two player game where opponents attempt to guess which character out of 24 possibilities their opponent has picked. Questions such as "Are you wearing glasses?" can help narrow down your choices. But choose your questions wisely and don't let your opponent name your character first!

Our main aim in this paper is to give the precise meaning to the concept of a "wise choice" of a question. To this end we shall introduce the concepts of discriminatory power of a polar question (section 3) and of a history of a question (section 4). These will allow to reformulate some known strategies of playing 'Guess Who?' in terms of our framework.

3 Discriminatory power of a polar question

We shall be using a rather traditional terminology concerning questions, following (Ajdukiewicz, 1974, p. 85–87). The schema of an answer to a question will be called

---

[2] http://www.boardgamecapital.com.

datum questionis. Thus in the case of a question "Which of the GW-characters is the chosen one?" the *datum questionis* is "x is the chosen one", where the range of x—the *uknown* of the question—is the set of all and only GW-characters.

We shall start with the concept of discriminatory power of a polar question. Let Ω stand for the set of all the GW-characters and let $\mathcal{G} \subseteq \Omega$. Let Q_{G_i} stand for a (polar) question concerning a certain property G_i: 'Does a GW-character possess the property G_i?'. By $\mathrm{d}Q_{G_i}$ we shall represent the (two-element) set of direct answers to Q_{G_i}. We shall represent the *positive extension* of G_i in $\mathcal{G}$ (that is, the set all the elements of $\mathcal{G}$ who possess the property G_i) by $^\mathcal{G}G_i^A$, and the *negative extension* of G_i in $\mathcal{G}$ (the set of all those characters who do not) by $^\mathcal{G}G_i^N$. By $\mid X \mid$ we shall represent the cardinality of the set X; thus $\mid \Omega \mid = 24$. We shall calculate *discriminatory power* of a question Q_{G_i} with respect to the set $\mathcal{G}$ in the following way:

$$^\mathcal{G}D_{G_i}^Q = \frac{min(\mid ^\mathcal{G}G_i^A \mid, \mid ^\mathcal{G}G_i^N \mid)}{max(\mid ^\mathcal{G}G_i^A \mid, \mid ^\mathcal{G}G_i^N \mid)} \times 100\% \qquad (1)$$

Thus discriminatory power of a question Q_{G_i} w.r.t. $\mathcal{G}$ is calculated as a fraction times 100 percent: of the numbers representing cardinality of the positive and negative extensions of G_i w.r.t. $\mathcal{G}$, where the smaller one goes to the numerator and the greater one goes to the denominator. As a result, the value of $^\mathcal{G}D_{G_i}^Q$ ranges from 0% (if either no character in $\mathcal{G}$ possesses the property G_i or all of them do, that is, either positive or negative extension of G_i is $\mathcal{G}$) to 100% (if the number of characters who possess G_i equals the number of characters who do not, that is, positive and negative extensions of G_i are equinumerous). We shall be rounding these values to the second decimal place. Occasionally we shall report the value of $^\mathcal{G}D_{G_i}^Q$ also as a fraction, as this gives information concerning the cardinalities of both positive and negative extensions of G_i in $\mathcal{G}$ as well as of $\mathcal{G}$ itself. We could omit reference to Q and define just discriminatory power of a property G_i w.r.t. $\mathcal{G}$. However, further on we want to be able to refer directly to questions.

The underlying rationale for the concept of discriminatory power of a polar question is that the higher its value, the more 'even' the split of a certain set of characters into two subsets, one consisting of those who posses certain property, the other of those who do not. The lower the value (but still greater than zero), the more 'lucky guess'-oriented is a question. If discriminatory power of a question equals zero then no answer to it narrows down the initial set of characters (to a non-empty proper subset).

The formula (1) does not allow to determine which of the two sets, $^\mathcal{G}G_i^A$ and $^\mathcal{G}G_i^N$, contains more elements. In general this information is not that important. Still, in reconstructions of real solutions to "Guess Who" tasks this issue may be of interest. We shall be using the following convention: if $\mid ^\mathcal{G}G_i^A \mid \leq \mid ^\mathcal{G}G_i^N \mid$, then $^\mathcal{G}D_{G_i}^Q = {}^\mathcal{G}D_{G_i}^{Q^A}$; otherwise $^\mathcal{G}D_{G_i}^Q = {}^\mathcal{G}D_{G_i}^{Q^N}$. Thus if the discriminatory power of a question Q_{G_i} w.r.t. the set $\mathcal{G}$ is symbolized by $^\mathcal{G}D_{G_i}^{Q^A}$ this means that the number of characters in $\mathcal{G}$ who possess the property G_i is at most equal to the number of those characters who do not; if it is symbolized by $^\mathcal{G}D_{G_i}^{Q^N}$ this means that there is more characters in $\mathcal{G}$ who possess G_i than those who do not.

Let us consider some examples. We shall start with calculating discriminatory power of a question Q_1: "Does the character wear spectacles?" w.r.t. the set Ω. As there are five characters who do, we have the following:

$$^{\Omega}D^{Q_1} = \frac{5}{19} \times 100\% = 26.32\% \qquad (2)$$

Notice, that in this case we calculated in fact $^{\Omega}D^{Q_1^A}$.

Now, let us consider the very same question with respect to the set $\mathcal{G}_1$ made up off all the characters from the first row; there is eight of them and only Claire wears spectacles. Thus:

$$^{\mathcal{G}_1}D^{Q_1} = \frac{1}{7} \times 100\% = 14.28\% \qquad (3)$$

Again, what we calculated is $^{\mathcal{G}_1}D^{Q_1^A}$.

Finally, let us consider the same question again, this time with respect to the set $\mathcal{G}_2$ made up of all the characters whose name begins with 'A'. There is four of them and none wears spectacles. As a result we have (this is again $^{\mathcal{G}_2}D^{Q_1^A}$):

$$^{\mathcal{G}_2}D^{Q_1} = \frac{0}{4} \times 100\% = 0\% \qquad (4)$$

In a sense, the questions with discriminatory power of 0% are unhelpful with respect to the set being considered. However, there are two cases of such lack of helpfulness. The first one is exemplified by the question Q_1 w.r.t. $\mathcal{G}_2$: in the whole space of possible solutions there are objects possessing this property, just none of them belongs to the subset being currently considered. Thus Q_1 *becomes* unhelpful when the set being considered is narrowed down from Ω (or $\mathcal{G}_1$, even) to $\mathcal{G}_2$. Notice, that the same pertains to the question Q_2: "Does the name of the character begin with 'A'?" w.r.t. $\mathcal{G}_2$, as all the considered object possess this property. The second case could be exemplified by a question Q_3: "Does the character smoke a pipe?", in which case there is no object in Ω with the desired property. Thus Q_3 is unhelpful w.r.t. Ω to begin with. Certainly, the difference between these two cases is a cognitive one and not a mathematical one. Moreover, there are good reasons to consider both types of 0% discriminatory power questions as a single class of questions, answers to which do not allow to narrow down the set being considered to some of its non-empty proper subsets. We shall go back to this issue later on.

Being not helpful is different, however, from being irrelevant. In case of questions like Q_4: "Is the character more than 6ft tall?" it is not possible to calculate its discriminatory power w.r.t. Ω, as there is not enough data to determine which characters possess this property and which do not. A question is unhelpful w.r.t. a certain set $\mathcal{G}$ being considered if either its positive or negative extension equals $\mathcal{G}$ (as we consider subsets of GW-characters the obvious condition of non-emptiness of $\mathcal{G}$ is tacitly assumed). A question is irrelevant if either its positive or negative extension cannot be determined. Both unhelpful and irrelevant questions are not forbidden in the game, however, asking them is certainly not "a wise choice".

It should be noted, that there are some similarities between the concept of discriminatory power of a polar question and the one of a bid of a question Q_{G_i} (Nica, 2016, p. 577), which in the case of 'Guess Who?' is the number of characters possessing the property G_i (or, in our terms, the cardinality of the positive extension of G_i). As the bids are defined in absolute numbers it is more difficult to compare bids of questions w.r.t. different pools of candidates; in case of our relative concept of discriminatory power this is quite straightforward. However, as we shall demonstrate later on, the stochastic analyses provided in Nica (2016) are valid for our framework as well.

4 History of a question

Now, let us consider an exemplary gameplay, from the subject 01 of our pilot study. The character chosen by the gamemaster is Eric.

Q_1 Are you a woman?
A_1 No.
Q_2 Are you less than 50 years old?
A_2 Yes.
Q_3 Do you have a beard?
A_3 No.
Q_4 Are you Bernard?
A_4 No.
Q_5 Do you have a moustache?
A_5 No.
Q_6 Do you have a cap?
A_6 Yes.
Q_7 Are you George?
A_7 No.
Q_8 Are you Eric?
A_8 Yes!

The range of the unknown of the question Q_1 is Ω. $^{\Omega}D^{Q_1}$, that is, discriminatory power of the question Q_1 w.r.t. Ω, equals 26.32% (that is, $\frac{5}{19}$). The range of the unknown of the question Q_2 is restricted to the negative extension of the property of being a woman (let us symbolize this property by G_1) and thus this is the set with respect to which we shall calculate discriminatory power of the question Q_2: $^{G_1^N}D^{Q_2} = 35.71\%$ (that is, $\frac{5}{14}$). Notice, that $^{\Omega}D^{Q_2}$, that is, discriminatory power of the question Q_2 w.r.t. Ω, equals 33,33% (that is, $\frac{6}{18}$). Taking into account consecutive questions and answers to them we can calculate discriminatory power of the question Q_7 w.r.t. the product of the: negative extension of the properties of being a woman, having a beard, having a moustache and being Bernard, and positive extensions of being less that 50 and having a cap to be equal to 100% ($\frac{1}{1}$). Finally, after removing George as the solution, discriminatory power of a question Q_8 w.r.t. the resulting set is 0%, as Eric is the only possible solution.

One important point is that actual 'Guess Who?' gameplays always start with a question concerning certain property of the GW-characters. However, in order to reconstruct the underlying reasoning we have to note that before this seeking for solution even starts a player needs to establish a conceptual frame for the problem to be solved. In other words, reasoning from an interpretation needs to be preceded by reasoning to an interpretation, by which the problem to be solved is defined (Stenning and van Lambalgen, 2008). In case of 'Guess Who' this may be framed as a question concerning which character has been chosen by the gamemaster (or the opponent): "Which exactly one x is such that x is the chosen one?". We shall be representing this question by Q_{in}. Employing Kubiński's numerical questions formalism (Kubiński (1980); see Wiśniewski (1995) for a short introduction) this question may be represented by the schema $1xTChO(x)$ (where $TChO$ stands for a one-place predicate 'is the chosen one'). Further on, polar questions will be represented, according to the set-of-answers methodology (Harrah, 2002; Peliš, 2016), as $?\{A, \neg A\}$.

In order to define the concept of a history of a question we need first to give precise meaning to the notion of a GW-gameplay (recall that Ω stands for the set of all the GW-characters).

Definition 4.1. *A GW-gameplay is a finite sequence* $\mathbf{g} = g_1, \ldots, g_t$ *such that:*

1. $g_1 = Q_{in}$;
2. $t > 1$ *is an odd number;*
3. *for each even number n such that $1 < n < t$: g_n is a question;*
4. *for each odd number o such that $2 < o \leq t$: g_o is an answer to g_{o-1};*
5. g_{t-1} *is a question of the form "Is the chosen one a?", where $a \in \Omega$ (in symbols: $?\{TChO(a), \neg TChO(a)\}$);*
6. g_t *is the affirmative answer to g_{t-1}.*

Thus a GW-gameplay (in a non-competitive setting) is a sequence starting with the initial question and then consisting of interwoven polar questions and answers to them. It ends with identification of the character chosen by the gamemaster.

History of a question is defined as a subsequence of a GW-gameplay:

Definition 4.2. *Let* $\mathbf{g} = g_1, \ldots, g_n$ *be a GW-gameplay and let g_m ($1 < m < n$) be a question in $\mathbf{g}$. History of g_m is a sequence* $\mathbf{h_{g_m}} = g_1, \ldots, g_{m-1}$.

We require that $m > 1$ as we are interested in histories of polar questions only and Q_{in}, being an initial question, does not have a history.

With this concept we are able to define the notion of the range of unknown of a question with respect to a certain history.

Definition 4.3. *Let Ω be the set of GW-characters and $\mathbf{h_Q}$ be a history of a question Q. Let:*

1. $Q_{G_i}, \ldots Q_{G_{k-1}}$ *be questions concerning properties $G_i, \ldots, G_{k-1}$ and such that in $\mathbf{h_Q}$ they are answered affirmatively, and*
2. $Q_{G_k}, \ldots Q_{G_l}$ *be questions concerning properties $G_k, \ldots, G_l$ and such that in $\mathbf{h_Q}$ they are answered negatively.*

The set which is the product of the affirmative extensions of $G_i, \ldots, G_{k-1}$ and the negative extensions of $Q_{G_k}, \ldots Q_{G_l}$ ($^\Omega G_i^A \cap \ldots ^\Omega G_{k-1}^A \cap {}^\Omega G_k^N \cap \ldots ^\Omega G_l^N$) is the range of unknown of Q w.r.t. the history $\mathbf{h_Q}$ (in symbols: $h_Q^\cap$).

Notice, that the set $h_Q^\cap$ is uniquely determined by $\mathbf{h_Q}$, but not *vice versa*.

Now, let us go back to the example from p. 4 and consider the history $\mathbf{h_{Q_4}}$ of a question Q_4 ("Are you Bernard?"), which is this:

> Which exactly one x is such that x is the chosen one? Are you a woman? No. Are you less than 50 years old? Yes. Do you have a beard? No.

Thus the range of unknown of Q_4 w.r.t. $\mathbf{h_{Q_4}}$ is the product of positive extension of being less than 50, negative extension of being a woman and negative extension of having a beard and it consists of 10 characters. Whether these extensions are determined w.r.t. to Ω or w.r.t. the ranges of unknown of the consecutive questions is not important, as both cases yield the same result.

5 'Guess Who?' and Inferential Erotetic Logic

Now let us turn to one possible normative yardstick by which the concept of a 'wise choice' of a question may be internationalized. Inferential Erotetic Logic (Wiśniewski, 1995, 2013a) is a framework for modeling reasoning in which questions play the roles of conclusions, and also of premises. The core of IEL are semantic relations of erotetic implication and of evocation of questions. They model formally the concepts: in the case of erotetic implication—of a question arising in view of some other question (and, possibly, some declarative evidence as well), in the case of evocation—of a question arising in view of a certain body of declarative information. We shall be interested in the former, as it may be employed in addressing the issue of processing consecutive questions in search for certain information—and this is exactly what cognitive activities of 'Guess Who?' players consist in. One important feature of IEL is that within this framework a logic of questions may be built on top on any logic of declaratives fulfilling some modest semantic criteria, characterized by Minimal Erotetic Semantics (Wiśniewski, 2013a, ch. 3 and 4). The most important of them is to assign truth values to formulas and to distinguish between the true and untrue ones.

We shall start with the definition of the relation of erotetic implication (Wiśniewski, 2013b, p. 67).

Definition 5.1. *A question Q erotetically implies (e-implies) a question Q^* on the basis of a set of d-wffs X (in symbols: $\mathbf{Im}(Q, X, Q^*)$) iff*

1. *for each $A \in \mathbf{d}Q$: $X \cup \{A\} \models \mathbf{d}Q^*$, and*
2. *for each $B \in \mathbf{d}Q^*$ there exists a non-empty proper subset Y of the set $\mathbf{d}Q$ such that $X \cup \{B\} \models Y$.*

If $X = \emptyset$, then we say that Q e-implies Q^* and we write $Im(Q, Q^*)$.

Thus e-implication meets the following conditions:

1. transmission of truth/soundness into soundness: if the question-premise is sound (i.e., there exists a true direct answer to this question) and all the declarative premises (if any) are true, then the question-conclusion is sound as well;
2. cognitive usefulness: each answer to the question-conclusion is useful in answering the question-premise (each answer to question-conclusion narrows down the class of possible answers to question-premise), provided that all the declarative premises (if any) are true.

The following theorem characterizes the relation between erotetic implication and discriminatory power of a polar question.

Theorem 1. *Let* $\mathbf{g} = g_1, \ldots, g_n$ *be a GW-gameplay and let* g_m $(1 < m < n)$ *be a question in* $\mathbf{g}$. *The question* g_m *is erotetically implied by the question* $g_1 = Q_{in}$ *on the basis of the set of answers to questions preceding* g_m *in* $\mathbf{g}$ *iff* $^{h^{\cap}_{g_m}}D^{g_m} > 0\%$, *that is, iff discriminatory power of the question* g_m *w.r.t. the range of unknown of* g_m *w.r.t. its history is greater that* 0%.

In slightly more intuitive terms the Theorem 1 states that g_m is erotetically implied by the initial question in a certain gameplay $\mathbf{g}$ on the basis of the set of answers to questions preceding g_m in $\mathbf{g}$ iff both positive and negative extensions of the property with which g_m is concerned are non-empty proper subsets of the range of unknown of g_m with respect to its history—that is, iff both positive and negative answers to g_m narrow down the set of objects being considered thus far. This is the gist of the second condition of the definition of erotetic implication. As for the first one we need to relay on the assumption that answers to all the polar question in the considered gameplay are given truthfully. If so, answers to consecutive questions in the gameplay eliminate objects which are not the chosen character, which, as a result, is located in consecutively narrowed down ranges of unknown.

In fact, if discriminatory power of g_m is as described in Theorem 1, then it is erotetically implied by Q_{in} on the basis of the empty set. This is a serious drawback of the IEL-based model of 'Guess Who?'. We shall go back to this issue in the section 6.

5.1 'Guess Who?' gameplays and erotetic derivations

Erotetic implication offers a formal basis for modeling question processing in 'Guess Who?'. The fact that this processing is systematic is captured by the concept of erotetic derivation (Wiśniewski, 2013b, p. 110–111).

Definition 5.2. *A finite sequence* $\mathbf{e} = \phi_1, \ldots, \phi_n$ *of wffs is an* erotetic derivation *of a direct answer A to a question Q on the basis of a set of d-wffs X iff* $\phi_1 = Q$, $\phi_n = A$ *and the following conditions hold:*

1. *for each question* ϕ_k *of* $\mathbf{e}$ *such that* $k > 1$:

a. $d\phi_k \neq dQ$, and
   b. ϕ_{k+1} is either a question or a direct answer to ϕ_k;

2. for each declarative formula ϕ_j of **e**:
   a. $\phi_j \in X$, or
   b. ϕ_j is a direct answer to ϕ_{j-1}, where $\phi_{j-1} \neq Q$, or
   c. ϕ_j is entailed by a certain set of d-wffs such that each element of this set precedes ϕ_j in **e**;

3. for each question ϕ_k of **e** such that $\phi_k \neq Q$: ϕ_k is erotetically implied by a certain question ϕ_j which precedes ϕ_k in **e** on the basis of the empty set, or on the basis of a set of d-wffs such that each element of this set precedes ϕ_k in **e**.

In IEL, the concept of erotetic derivation is in fact an auxiliary one, by which erotetic search scenarios are defined (Wiśniewski, 2003). Intuitively, erotetic derivations form distinct paths of a scenario for a given question and different paths are determined by different answers to auxiliary questions. A scenario shows which answer to an initial question may hold given different answers to auxiliary questions obtained. It is possible to represent the space of solutions in 'Guess Who?' in a form of such a scenario (a nice example of such a representation which may be readily transformed into a scenario can be found here: http://chalkdustmagazine.com/blog/cracking-guess-board-game/). This may be useful, in particular, while deciding on *a priori* optimal strategy in the game. However, we are interested in single GW-gameplays, for which erotetic derivations generally are adequate formal counterparts.

Theorem 2. *Let* $\mathbf{g} = g_1, \ldots, g_n$ *be a GW-gameplay. The sequence* $\mathbf{g}$ *is an erotetic derivation of a direct answer* g_n *to the question* $g_1 = Q_{in}$ *iff for each question* g_m *in this sequence* $^{h_{g_m}^{\cap}}D^{g_m} > 0\%$, *that is, iff discriminatory power of each question* g_m *w.r.t. the range of unknown of* g_m *w.r.t. its history is greater that* 0%.

There are some uncontroversial cases of GW-gameplays which are erotetic derivations of answers to Q_{in} being their last elements. However, there is at least one tricky case. Consider a GW-gameplay $\mathbf{g} = g_1, \ldots, g_n$ such that g_{n-1}, the last question in $\mathbf{g}$ is a rhetorical one—that is, such that the range of unknown of g_{n-1} w.r.t. its history consists of just one character; our exemplary gameplay (p. 4) is exactly such. As a result, discriminatory power of g_{n-1} is 0% and this question is not erotetically implied by Q_{in} on the basis of the set of answers to preceding questions (cf. Theorem 1). Still, $\mathbf{g}$ is a legitimate GW-gameplay, as the character chosen needs to be identified by the positive answer to a question. Let us call such GW-gameplay in which last question is a rhetorical one a *complete* GW-gameplay. We may prove the following theorem:

Theorem 3. *Let* $\mathbf{g} = g_1, \ldots, g_n$ *be a complete GW-gameplay. The sequence* $\mathbf{g}' = g_1, \ldots, g_{n-2}, g_n$ *is an erotetic derivation of a direct answer* g_n *to the question* $g_1 = Q_{in}$ *iff for each question* g_m *in this sequence* $^{h_{g_m}^{\cap}}D^{g_m} > 0\%$, *that is, iff discriminatory power of each question* g_m *w.r.t. the range of unknown of* g_m *w.r.t. its history is greater that* 0%.

The sequence **g'** differs from **g** in that from **g'** the last question was removed, that is, the rhetorical g_{n-1}. Because the range of unknown of g_{n-1} w.r.t. its history in **g** is a singleton set this means that g_n—the final claim of identity of the chosen character—is entailed by the set of declarative formulas from this history. As all these formulas belong to **g'** as well, g_n is entailed by the set of declaratives which precede g_n in **g'**. As a result, **g'** is an erotetic derivation of a direct answer g_n to the question $g_1 = Q_{in}$. Notice however, that **g'** no longer is a GW-gameplay (cf. Definition 4.1), as in GW-gameplay each declarative formula must be preceded by a relevant question and both g_{n-2} and g_n are declarative formulas.

6 'Guess Who?' and situational semantics

IEL offers a normative model for 'Guess Who' gameplays in terms of erotetic implication. One problem is, that under this normative account all the auxiliary questions are in general erotetically implied by the initial one only; what holds between the auxiliary ones is, at best, the relation of weak erotetic implication (Urbański et al, 2016). As a result, all the dynamics of information stemming from asking and answering auxiliary questions is explicitly expressed only by declarative information in the form of changing ranges of unknown. Now let us turn to a model which is erotetically more informative in that it accounts for more fine-grained relations which hold between auxiliary questions. Our framework is an extension of a version of situational semantics (Wiśniewski, 2013a), which we previously applied in modeling inferential question processing involved in solutions to a specific class of abductive problems (Urbański and Żyluk, 2018). Also in (Urbański and Żyluk, 2018) the interested Reader will find more detailed justification of the choice of our formal tools.

There are two fundamental concepts on which this model is based. The first one, the concept of situation, we shall consider as a primitive one, following in that Keith Devlin's (1991, p. 70) claim that "situations are just that: situations". The second one is the concept of a topic. Interpreting it in terms of situational semantics we shall follow some general lines proposed by Jan Van Kuppevelt, according to whom "[t]he term topic (...) refer[s] to a topic notion which concerns the 'aboutness' of (sets of) utterances" (Van Kuppevelt, 1995, p. 111). As the author claims:

> The notion presupposes that a discourse unit U—a sentence or a larger part of a discourse—has the property of being, in some sense, directed at a selected set of discourse entities (a set of persons, objects, places, times, reasons, consequences, actions, events or some other set), and not diffusely at all discourse entities that are introduced or implied by U. This selected set of entities in focus of attention is what U is about and is called the topic of U. (Van Kuppevelt, 1995, p. 112)

The following definition of a situational model of the Classical Propositional Calculus language is just to witness that there is nothing particularly non-standard in the assumed semantics (Wiśniewski, 2013a, p. 39). It will be needed in order to introduce the concept of relevance (*Form$_{CPC}$* stands for the set of CPC formulas).

Definition 6.1. *A* situational model *of the CPC language is an ordered pair* $\mathbb{M} = \langle U, v \rangle$, *where U is a non-empty set (the universe of* $\mathbb{M}$*) and v is a function Form$_{CPC}$* $\mapsto$ 2^U *such that:*

1. *for each propositional variable p_i, $v(p_i) \subseteq U$*
2. *for each $A, B \in Form_{CPC}$:*

   a. $v(\neg A) = U - v(A)$,
   b. $v(A \wedge B) = v(A) \cap v(B)$,
   c. $v(A \vee B) = v(A) \cup v(B)$,
   d. $v(A \to B) = v(\neg A) \cup v(B)$,
   e. $v(A \leftrightarrow B) = (v(\neg A) \cup v(B)) \cap (v(\neg B) \cup v(A))$.

We shall be considering topics w.r.t. (or in) some situational model. Under this account, a topic Γ may be interpreted as a subset of the model's universe ($\Gamma \subseteq U$). However, it may happen that some information, expressed by a certain formula is not relevant—in a very intuitive sense, as yet—to a certain topic. This is why we need an additional concept of a relevance model.

Definition 6.2. *Let* $\mathbb{M} = \langle U, v \rangle$ *be a situational model and* $\Gamma \subseteq U$ *be a topic in* $\mathbb{M}$. *A* situational relevance model *of Γ w.r.t.* $\mathbb{M}$ *is an ordered triple* $\mathbb{N} = \langle \Gamma, w, \mathbb{M} \rangle$, *where w: Form$_{CPC}$* $\nrightarrow$ 2^Γ *is such that:* (*) *if $w(A) \subseteq \Gamma$, then $w(A) = v(A)$.*

Notice, that w is just a partial function from *Form$_{CPC}$* to 2^Γ, thus for some formulas no subset of Γ may be assigned. Further on we shall refer to such models simply as to relevance models.

As a result we obtain a simple and straightforward characteristics of topic relevance:

Definition 6.3. *A topic Γ' is* relevant *to a topic Γ w.r.t.* $\mathbb{N}$ *iff $\Gamma' \subseteq \Gamma$.*

Thus a formula A is relevant to a topic Γ w.r.t. $\mathbb{N}$ iff $w(A) \subseteq \Gamma$. The condition (*) imposed on w in definition of relevance model may be expressed as follows: (*') If A is relevant w.r.t. $\mathbb{N}$, then $w(A) = v(A)$.

Admittedly, definition 6.3 imposes a rather strong condition on the concept of relevance. Some weakening is possible, as, for example, in the following form:

Definition 6.4. *A topic Γ' is* somewhat relevant *to a topic Γ w.r.t.* $\mathbb{N}$ *iff $\Gamma' \cap \Gamma \neq \emptyset$.*

However, in what follows we shall employ the stronger version.

6.1 Sifting and funneling

The basic intuition underlying the concepts of sifting and funneling may be convincingly exemplified with a help of a simple *20 questions* gameplay (this is a game in which the player's task is to identify what kind of objects the gamemaster has in mind, by asking polar questions; see Urbański and Żyluk 2018, p. 866).

Q1 Is it an animal?
A1 Yes
Q2 Is it a mammal?
A2 Yes
Q3 Is it a rodent?
A3 Yes
Q4 Is it a rat?
A4 No
Q5 Is it a pet?
A5 Yes
Q6 Is it a guinea pig?
A6 No
Q7 Is it a hamster?
A7 Yes

Consider the sequence of questions Q_1, Q_2, Q_3, Q_4. Each consecutive question funnels (in view of affirmative answer to the preceding one) the set of objects to be considered. The role of Q_5, in view of the negative answer to Q_4, is to funnel Q_3, in a sense, in a different direction. Now, consider the questions Q_6, Q_7. They sift the set of pet rodents in order to identify the solution. These are relatively simple examples. The relations of sifting and funneling come in many flavours and we shall employ at least some of them in our models of GW-gameplays.

Definitions of sifting and funneling employ logic of questions, situational semantics, and topic relevance (we refer the Reader to Urbański and Żyluk (2018) for more detailed account). We shall use the following notation: Let A^*, where A is a CPC formula, stand for any of the following two: A, $\neg A$. Let $\bar{A}$ and $\bar{\bar{A}}$ stand for a pair of *complementary formulas based on* A: if $\bar{A} = A$, then $\bar{\bar{A}} = \neg A$, and if $\bar{\bar{A}} = A$, then $\bar{A} = \neg A$. Thus, in case of yes-no question $Q = ?\{A, \neg A\}$, Q has complementary formulas based on A as its direct answers.

Definition 6.5. *The questions* $Q_1 = ?\{A_1, \neg A_1\}$, ..., $Q_n = ?\{A_n, \neg A_n\}$ *are sifting questions w.r.t. a topic* Γ *of a certain relevance model* $\mathbb{N}$ *iff for every i and j ($1 \leq i, j \leq n$): $w(A_i^*)$ is non-empty, for each A_i at least one of the $w(\bar{A}_i), w(\bar{\bar{A}}_i)$ is a proper subset of Γ, and*

1. **Disjoint sifting:** *if* $A_i \neq A_j$, *then* $w(A_i^*) \cap w(A_j^*) = \varnothing$ *for at least one of the* $\bar{A}_i, \bar{\bar{A}}_i$ *and at least one of the* $\bar{A}_j, \bar{\bar{A}}_j$.
2. **No-common-core sifting:** $w(A_1^*) \cap ... \cap w(A_n^*) = \varnothing$ *for at least one element of each of the pairs* $\bar{A}_i, \bar{\bar{A}}_i$.
3. **Something's-left-behind sifting:** *Let* A_i^* *be a certain answer to* Q_i *and* A_j^* *be a certain answer to* Q_j; *if* $A_i \neq A_j$, *then both* $w(A_i^*) - w(A_j^*) \neq \varnothing$ *and* $w(A_j^*) - w(A_i^*) \neq \varnothing$.

Thus $?\{A_1, \neg A_1\}$, ..., $?\{A_n, \neg A_n\}$ are sifting w.r.t. a topic Γ of a certain relevance model $\mathbb{N}$ iff sets of situations assigned to answers to these questions are non-empty, for each question there exists at least one answer such that set of situation assigned to it is a proper subset of Γ and:

1. in case of disjoint sifting: there is a selection of answers to these questions such that sets of situations assigned to these answers are pairwise disjoint; in other words, the selected sets $w(A_1^*)$, ..., $w(A_n^*)$ are partitioning Γ, albeit this partitioning need not to be exhaustive;
2. in case of no-common-core sifting: there is a selection of answers to these questions such that the product of all the selected $w(A_1^*)$, ..., $w(A_n^*)$ is the empty set, that is, there is no situation common to the claims made by all these sentences;
3. in case of something's-left-behind sifting: for each pair of distinct questions, distinct answers to them are such that there are situations which are covered by one of them and not by the other.

In the case of funneling the general idea may be expressed in terms of this relation holding between two topics: a topic Γ_1 *funnels* a topic Γ_2 iff $\Gamma_1 \subset \Gamma_2$, and both are non-empty. However, we shall be interested in funneling holding between a polar question and a topic, and between two questions.

Definition 6.6. *A question $Q = ?\{A, \neg A\}$ funnels a topic Γ of a certain relevance model $\mathbb{N}$ iff for some answer A^* to Q: $w(A^*) \subset \Gamma$, and both are non-empty.*

Thus a question $?\{A, \neg A\}$ funnels a topic Γ w.r.t. a certain relevance model $\mathbb{N}$ iff one of the answers to $?A$, such that a set of situations assigned to it is non-empty, narrows down Γ.

The concept of funneling may be easily generalized to a relation which holds between two questions of any type:

Definition 6.7. *A question $Q_1 = ?\{A_1, \ldots, A_n\}$ funnels a question $Q_2 = ?\{B_1, \ldots, B_m\}$ w.r.t. a certain relevance model $\mathbb{N}$ iff for each A_i ($1 \leq i \leq n$): $w(A_i) \subset w(dQ_2)$, and both are non-empty.*

At least two specific kinds of funneling are worth mentioning: regular and weakly regular ones.

Definition 6.8. *A question $Q_1 = ?\{A_1, \ldots, A_n\}$ regularly funnels a question $Q_2 = ?\{B_1, \ldots, B_m\}$ w.r.t. a certain relevance model $\mathbb{N}$ iff for each A_i ($1 \leq i \leq n$) there exists B_j ($1 \leq i \leq m$) such that $w(A_i) \subset w(B_j)$, and both are non-empty.*

Thus Q_1 regularly funnels Q_2 w.r.t. a certain relevance model $\mathbb{N}$ iff for each answer A_i to Q_1 there exists an answer B_j to Q_2 such that the topic $w(A_i)$ funnels the topic $w(B_j)$. As a result, the question 'Is John a logician or is he a philosopher?' (construed as a question with just two direct answers: 'John is a logician', 'John is a philosopher') is regularly funneled by the question 'Is John a modal logician or is he an analytic philosopher?' (with similar proviso). Notice, that we do not require sets of situations assigned to answers to neither Q_1 nor Q_2 to be disjoint.

Definition 6.9. *A question $Q_1 = ?\{A_1, \ldots, A_n\}$ weakly regularly funnels a question $Q_2 = ?\{B_1, \ldots, B_m\}$ w.r.t. a certain relevance model $\mathbb{N}$ iff for some A_i ($1 \leq i \leq n$) there exists B_j ($1 \leq i \leq m$) such that $w(A_i) \subset w(B_j)$, and both are non-empty.*

The question 'Is John a logician or is he a philosopher?' is weakly regularly funneled by the question 'Is John a modal logician or is he just showing off?': for weakly regular funneling it is enough that topic assigned to one of the answers to Q_1 funnels a topic assigned to one of the answers to Q_2.

For our purposes we shall use a version of weak regular funneling restricted to simple yes-no questions. Thus a question $Q_i = ?\{A_i, \neg A_i\}$ weakly regularly funnels a question $Q_j = ?\{A_j, \neg A_j\}$ w.r.t. a certain relevance model $\mathbb{N}$ iff for some answer A_i^* to Q_i and some answer A_j^* to Q_j: $w(A_i^*) \subset w(A_j^*)$, and both are non-empty.

Now let us have a look on how these relations work in the case of 'Guess Who?' We shall start with our exemplary gameplay (see p. 4) and go through it question by question.

1. Q_{in}: $1xTChO(x)$, "Who is the chosen one?"
   - Q_{in} establishes Ω, the set of GW-characters, as the universe. (One sidenote is in order here: strictly speaking, the universe consists of situations, not characters. So, its elements are situations like 'Claire is the chosen one', etc. However, as there is no real danger of an ambiguity, we shall be using just names as signifiers for them.)

2. Q_1: "Are you a woman?"
   - Q_1 funnels Ω.

3. Q_2: "Are you below 50?"
   - Q_2 funnels both Ω and negative extension of being a woman in Ω.
   - Q_2 funnels Q_1 and funnels it weakly regularly, but not regularly.

4. Q_3: "Do you have a beard?"
   - Q_3 funnels Ω, Q_1, Q_2; Q_3 weakly regularly funnels Q_1, Q_2; Q_3 does not regularly funnel them.

5. Q_1, Q_2, Q_3 are sifting questions, and this is disjoint sifting, because of the following:
   - no woman has a beard or is over 50,
   - no man over 50 has a beard.

6. Q_4: "Are you Bernard?"
   - Q_4 funnels Ω and all the previous questions.

7. Q_5: "Do you have a moustache?"
   - Q_5 funnels Q_4, and funnels it weakly regularly. $Q_1, \ldots, Q_5$ are sifting questions, but just of the no-common-core type.

8. Q_6: "Do you have a cap?", Q_7: "Are you George?"
   - Similar as in the case of Q_5.

9. Q_8: "Are you Eric?"

- This question does not funnel the topic established by its history, and there is no sifting either (notice, that the negative extension of the property being asked w.r.t. the history of Q_8 is the empty set; it is a rhetorical question and its discriminatory power at this stage of the gameplay is 0%).

Notice, that each of the questions $Q_1, \ldots, Q_8$ is e-implied by Q_{in} but between $Q_1, \ldots, Q_8$ there are no e-implications. Also, Q_8 is not e-implied by Q_{in} on the basis of the answers to $Q_1, \ldots, Q_7$: the negative answer to Q_8 does not narrow down the set of answers to Q_{in} to a non-empty subset.

Now let us turn to an example in which some funneling and sifting hold, but there is no e-implication.

1. Q_{in}: $1xTChO(x)$, "Who is the chosen one?"

   - Q_{in} establishes Ω as the universe.

2. Q_1: "Are you blond(e)?"
3. Q_2: "Are you male?"

   - Both Q_1, Q_2 funnel Ω.
   - Q_1 funnels Q_2 (w.r.t. Ω) and *vice versa* (as both establish a logical division of Ω).
   - No regular or weak regular funneling.
   - Q_1, Q_2 are sifting questions, but just of the 'something's-left-behind' type.
   - Notice, that there is no e-implication between Q_1 and Q_2.

We can establish some relation between sifting, funneling and discriminatory power.

Theorem 4. *If discriminatory power of a question Q w.r.t. a set $\mathcal{G}$ is 0% ($^{\mathcal{G}}D^Q = 0\%$), then Q does not funnel $\mathcal{G}$ (the reverse implication does not hold). In such case Q would not be a sifting question w.r.t. $\mathcal{G}$ either.*

As for the relation between the IEL model and the situational model, there are two rather obvious cases:

1. If a question Q_1 regularly funnels a question Q_2, then Q_2 e-implies Q_1.
2. If a question Q_1 weakly regularly funnels a question Q_2, then Q_2 weakly e-implies Q_1.

The reverse implications do not hold. Funneling itself does not guarantee any kind of e-implication, nor is sifting. Thus sifting and funneling are more fine-grained relations that e-implications in that they allow to account for some semantic connections between questions the e-implications are unable to grasp.

7 Conclusion

First, and unsurprising, conclusion is that for a question in 'Guess Who?' in order to be a 'wise choice' its discriminatory power has to be above 0%, w.r.t. the range of unknown being considered. The second one is, that bold and cautious moves in playing 'Guess Who?' may be characterized in terms of discriminatory power: the lower the discriminatory power of a question (but still greater than 0%), the bolder move it forms, the higher the discriminatory power—the more cautious is the move. Thus in the competitive setting (Nica, 2016, p. 577) the 'low risk play', advisable when a player has an upper hand, consists in asking question with the discriminatory power as close to 100% as possible. Questions with discriminatory power lower than that are 'catching up moves', advisable when a player is in the weeds; the actual boldness of the move depends on the deep of those weeds, that is, on how many GW-characters a player is behind the opponent (Nica (2016) offers an appropriate formula). The same holds for playing strategies: the binary choice strategy (Nica 2016, p. 577; see also http://www.geekyhobbies.com/how-to-win-guess-who-within-six-turns/ and http://chalkdustmagazine.com/blog/cracking-guess-board-game/) may be characterized as consecutively choosing the questions with the highest discriminatory power, w.r.t. changing range of unknown. As Nica (2016, p. 578) nicely summarizes it:

> In terms of these features, 'Guess Who?' can be thought of as a toy model of some situations where two firms are competing to be the first to bring a product to market and have the ability to manage risk vs reward on the time it takes for product development. Our analysis suggests that the leading firm should take no risks (i.e. try only to minimize expected development time) and the trailing firm should take the smallest risk possible catch up to the leading firm.

Some directions for further studies may be indicated. One interesting issue is possible generalization of sifting to types of questions other than polar ones. Another is application of our models to the tasks and games other than the ones mentioned in the paper. For example, in case of Wason Selection Task construed in an interrogative setting (that is, in which questions asked in order to solve the task are made explicit) the informatively correct solution is modeled by an erotetic search scenario, in which questions asked are e-implied by the initial one and declarative evidence, while in case of suboptimal solutions the questions are only weakly e-implied at best. Thorough comparison of our frameworks, IEL one and situational one, and their efficacy in modeling human reasoning, is the final direction of further investigations we would like to point at.

References

Ajdukiewicz K (1974) Pragmatic Logic. D. Reidel Publishing Company Dordrecht & PWN, Dordrecht & PWN

Devlin K (1991) Logic and information. Cambridge University Press

Harrah D (2002) The logic of questions. In: Handbook of philosophical logic, Springer, pp 1–60

Kubiński T (1980) An Outline of the Logical Theory of Questions. Akademie-Verlag, Berlin

Nica M (2016) Optimal strategy in "Guess Who?": Beyond binary search. Probability in the Engineering and Informational Sciences 30(4):576–592

Peliš M (2016) Inferences with Ignorance: Logics of Questions. Charles University in Prague, Karolinum Press

Stenning K, van Lambalgen M (2008) Human reasoning and cognitive science. MIT Press, Cambridbe, MA

Urbański M, Żyluk N, Paluszkiewicz K, Urbańska J (2016) A formal model of erotetic reasoning in solving somewhat ill-defined problems. In: Mohammed D, Lewiński M (eds) Argumentation and Reasoned Action. Proceedings of the 1st European Conference on Argumentation, Lisbon 2015, College Publications, London, vol II, pp 973–983

Urbański M, Żyluk N (2018) Sets of situations, topics, and question relevance. In: Oswald S (ed) Proceedings of the 2nd European Conference on Argumentation (to appear), College Publications, London

Van Kuppevelt J (1995) Discourse structure, topicality and questioning. Journal of linguistics 31(1):109–147

Wiśniewski A (1995) The posing of questions: Logical foundations of erotetic inferences. Kluwer Academic Publishers, Dordrecht–Boston–London

Wiśniewski A (2003) Erotetic search scenarios. Synthese 134(3):389–427

Wiśniewski A (2013a) Logic and sets of situations. In: Essays in Logical Philosophy, LiT Verlag, Berlin, pp 33–46

Wiśniewski A (2013b) Questions, Inferences, and Scenarios. College Publications, London

Natural Deduction Method for Solving CL-based Puzzles

Agata Tomczyk

Abstract Combinatory logic (CL) was one of the first formal systems which were focused on the notion of function and function abstraction. CL and other systems (like λ-calculus), though developed independently, are on many levels similar. Properties of these systems are also universal—they can be found *e.g.* in numerous functional programming languages and different logical systems, for instance in natural deduction (ND). It was later discovered that the rules proposed by Gentzen for ND and those by Church for simply typed λ-calculus are comparable—*e.g.* normalization in ND corresponds to β-reduction in λ-calculus. Additionally, CL expressions can be transformed to λ ones and *vice versa*. Due to this interchangeability of CL and λ expressions, there are numerous ways of solving Smullyan's CL-based puzzles. On the basis of these similarities we propose a formal way of solving Smullyan's CL-based puzzles, by way of Curry-Howard isomorphism. In this paper the said approach will be formalized in ND-style. The presented method uses unmodified and type-free combinators. Solving a puzzle is understood as creating a proof for a given formula, starting with a set of pre-assumed conditions. Proofs for a given puzzle are thus represented by trees, in which each node is obtained by applying one of the mentioned rules. The application of the rules is shown while solving few selected Smullyan's puzzles.

Key words: combinatory logic, combinator, reduction, ND, puzzle solving

---

Agata Tomczyk
Department of Logic and Cognitive Science
Faculty of Psychology and Cognitive Science
Adam Mickiewicz University, Poznań, Poland
e-mail: a.tomczyk@protonmail.com

1 Introduction

Raymond Smullyan was a man of many talents. Aside from being a mathematician, logician, musician and magician, he had a good understanding of human fear. Particularly about the fear of formal sciences. Therefore, he introduced many puzzles based on different formal systems, but written in a laymen-friendly manner—descriptive and simple. We can distinguish two groups of the puzzles (Kolany, 1996). The first group consists of puzzles that refer to the classical logic and the idea of the bivalence[1]. For instance, a series of *Knights and Knaves* puzzles can be easily solved using truth tables and a number of Boolean laws. Within the second group we can find puzzles explaining Gödel's incompleteness theorem and recursive functions. In his book dedicated to Haskell Curry entitled *To Mock a Mockingbird*, Smullyan presented a number of puzzles based on combinatory logic (Smullyan, 2012). In order to explain the notion of a combinator, Smullyan uses a number of bird metaphors. The choice of leitmotiv of the said puzzles is due to the Curry's great passion—bird watching. Thus birds in Smullyan's world are functions—by means of which Smullyan describes certain combinators and their properties. The use of natural language provides an insight into a structure of an algorithm that could be formalized. Furthermore, thanks to the notion of "propositions as types", a number of different corresponding systems can be introduced for solving Smullyan's puzzles by means of isomorphic mechanisms.

In this paper we present a simple and formal approach to solving Smullyan's combinatory logic (hereafter: CL) based puzzles, by adapting natural deduction-style (ND) method. The choice of this specific method is due to Smullyan's descriptive manner. By assuming that certain conditions hold, we are able to reach the conclusion by employing a set of rules based on deduction laws of CL. In the first part of the paper a meta-theoretical background of combinatory logic is given. On the basis of it we introduce a ND-style system of CL; such system is adapted to solving CL-based puzzles. As the proposed method is similar to proof procedures, the notion of resolved puzzle is reduced to the adapted definition of a proof.

A number of strategies for solving Smullyan's and Smullyan-inspired puzzles had already been presented *e.g.* as computer programs. Most often, Prolog is the programming language of choice. Such implementations have been proposed by Jerardi (1987), Casimir (1987) and Ramsey (1986). Other ways of solving Smullyan's puzzles have been developed on the basis of automated reasoning. Aszalos (2002) and Kolany (1996) both use Classical Propositional Calculus (CPC) syntax, along with truth-tables. However, all of the above methods refer to the first group of Smullyan's puzzles. The notion of two-valued logic as the base of these puzzles allows us to utilize numerous proof methods of CPC. The second group of Smullyan's puzzles has not been as widely examined in terms of solving strategies. Lark combinator is often the main interest, due to its properties. Statman (1989) examines how the problem of equality of two applicative Lark combinators is solvable, while Sprenger and Wymann-Böni (1993) provide a decision algorithm for this particular combinator. Interesting approach has

---

[1] We can distinguish puzzles utilizing both Classical Propositional Calculus as well as First-Order Logic.

been brought up by Keenan (2014), who introduced a graph-based method, combined with λ-calculus, in order to solve Smullyan's CL-based puzzles. Strategy presented in this paper, similarly to these mentioned above, also employs a logical system and can become a base for an algorithm. As CL had later on become a reference point for many programming languages, such transition seems natural. Functional programming languages with well-optimized recursion mechanism could allow to implement such system without a need for many structural changes.

2 Combinatory logic

In the 1920s and 1930s two formal systems were introduced, both based on the notion of function and function abstraction. These concepts were not anything new back then—they have been already discussed by Gottlob Frege and Bertrand Russell, however no definition has been provided at the time (Hindley and Cardone, 2006). A formalization was needed, so the authors of two new systems aimed to define certain properties and operations on functions. The first system, developed in 1926 by Moses Schönfinkel, was combinatory logic. Roughly at about the same time, another logician and mathematician, Haskell Curry, independently started working on simplifying the process of substitution. He discovered works of Schönfinkel in 1927 and focused on combinatory logic. His doctoral thesis *Grunladen der Kombinatorischen Logik* was an introduction to formal representation of building terms from variables and constants. However, the original combinatory logic system, along with the second one, λ-calculus, was proved to be inconsistent (due to arbitrary fixed points) (Sorensen and Urzyczyn, 2006). In order to provide a consistent system built on a notion of a pure function, we can distinguish a number of different subsystems.

Overall, by combinatory logic we can define a logical theory built around a notion of function and recursion. These systems became widely popular and are now connected to many other areas of formal sciences. Despite their diverse application, the motivations for its invention are rather alike. The main goal was to introduce minimalistic system, which could become an universal computational model. In order to achieve that, one of the problems had to be tackled—namely the process of substitution. In most formal systems the problems generated by substitution are present due to the presence of the bound variables. Combinatory logic provides very straight-forward solution regarding this issue. There are no bound variables found in its terms. As a result we can notice a reduction of the number of substitution restrictions. This property made combinatory logic a very useful tool in other disciplines, for example mathematics or computer science (Wolfengagen, 2003). Its simple yet comprehensive and expressive syntax became a base for many programming languages. This alone allowed to brought up the topic of correspondence between such languages and certain formal systems.

2.1 Curry-Howard Isomorphism

Curry-Howard Isomorphism (also known as "propositions as types") came to life due to observations made by Haskell Curry and William Howard. Curry noticed that types of given combinators can be also thought of as formulae of intuitionistic logic (Wadler, 2015). For example, formula of the form ⌜$A \rightarrow B$⌝ can be read as a type of function which takes the argument of type A and returns value of a type B. Later on, combinatory logic has been proven to correspond to other system—namely Hilbert-style axiomatic system. Howard noticed similar correspondence between λ-calculus and ND (for intuitionistic logic). Simply put, "propositions as types" theory describes a correspondence between a given logic system and λ-calculus or CL. Within a theory we also examine the multilevel aspect of the isomorphic correspondence. For example, we can study how for each proposition in a considered logic system an equivalent programming language type can be found. Going deeper, a proof for such proposition would be corresponding to program of that type. And, finally, simplifying such proof is corresponding with evaluating a program of a type connected to this proposition (Wadler, 2015). As discussed above, we can distinguish two main pairs of systems described within Curry Howard Isomorphism. First model connects different formulations of λ-calculus structure with ND and the second one—combinatory logic with Hilbert style deduction system. Additionally, λ-expressions can be translated into combinatory logic terms and *vice versa*. We can define a function $\mathcal{T} : \lambda \rightarrow T$ translating any λ-expression to CL term (Sorensen and Urzyczyn, 2006):

Definition 2.1.

$$\mathcal{T}(x) = x;$$
$$\mathcal{T}(\lambda x.x) = I;$$
$$\mathcal{T}(\lambda xy.x) = K;$$
$$\mathcal{T}(\lambda xyz.xz(yz)) = S;$$
$$\mathcal{T}(MN) = \mathcal{T}(M)\mathcal{T}(N)$$

where M and N are two arbitrary λ-expressions.

The inverted function can be defined accordingly. Having the above function defined, we can easily combine CL with ND method. Therefore adapting descriptive way of solving Smullyan's CL puzzles to ND can be done without causing any harm, as both systems are connected due to substantial similarities (through λ-calculus).

2.2 Syntax

One of the major motivations behind developing CL was minimizing the set of first-order logic's set of constants. As a result, CL's language became minimalistic with only a specified number of constants, depending on considered subsystem of CL. The most basic subsystem is based on 3 combinators: I, K, S, which have their equivalence in CPC:

(I) : I P = P	$A \to A$
(K): K P Q = P	$A \to (B \to A)$
(S) : S P Q R = P R (Q R)	$(A \to (B \to C)) \to ((A \to B) \to (A \to C))$

Therefore the language of $I - K - S$ subsystem of CL can be defined. The below definitions are based on ones found in Sorensen and Urzyczyn (2006).

Definition 2.2. *The set of well formed formulas of CL is defined with respect to the following grammar:*

$$T ::= x \mid I \mid K \mid S \mid (TT)$$

where x is an element of the infinite set of variables $V = \{x_0, x_1, x_2, \cdots\}$ *and T is called a term of CL.*

Having specified the grammar, in order to define the substitution process we have to introduce the process of distinguishing the free variables of CL terms.

Definition 2.3. *The set FV of free variables of a given term T (denoted by $FV(T)$) is obtained by way of induction:*

$$FV(x) = \{x\};$$
$$FV(I) = \{\};$$
$$FV(K) = \{\};$$
$$FV(S) = \{\};$$
$$FV(T_iT_j) = FV(T_i) \cup FV(T_j).$$

Definition 2.4. *A substitution of CL term T_j for variable $x_i \in V$ is defined as follows:*

$$x_k[x_i/T_j] = \begin{cases} x_k, & \text{if } k \neq i, \\ T_j, & \text{if } k = i, \end{cases}$$
$$I[x_i/T_j] = I;$$
$$K[x_i/T_j] = K;$$
$$S[x_i/T_j] = S;$$
$$(T_0T_1)[x_i/T_j] = T_0[x_i/T_j]T_1[x_i/T_j].$$

In order to obtain more combinators (higher-order function of the form $P = Q$, where P and Q are arbitrary terms of CL consisting of other functions) we can introduce a set of deduction rules[2] (Barendregt, 2013):

(i) $P = P$;
(ii) If $P = Q$ then $Q = P$;
(iii) If $P = Q$ and $Q = R$ then $P = R$;
(iv) If $P = P'$ then $PR = P'R$;
(v) If $P = P'$ then $RP = RP'$.

The above set of rules defines certain properties that can be used—identity, commutativity and transitivity. Additionally, if we were to add the same argument or function

---

[2] These rules, along with I, K, S combinators can be also used to define a reduction relation and its properties—namely commutativity and transitivity (Sorensen and Urzyczyn, 2006).

on both sides of the resulting equation, terms appearing on both sides of it would still be equal. We can use these rules to prove that a given combinator can be reduced to another one. For example, let $M = SII$. We can prove that it can be reduced to $Mx = xx$, using the combinators introduced in the last section:

$$\begin{aligned} Mx &= SIIx \\ &= Ix(Ix) \\ &= xx. \end{aligned}$$

3 Smullyan's puzzles

This section consists of the description of modified ND system for solving CL-based puzzles. It is based on the CL system defined in the preceding section. The structure of trees representing the solution of a given puzzle has been determined by descriptive manner of Smullyan's method.

3.1 Pre-assumed conditions

In *To Mock a Mockingbird* the difficulty of puzzles increases as we proceed. A set of introductory puzzles is a relatively easy one. Smullyan starts with the notion of combinators through bird metaphors, which are embedded into a compelling storyline. In "a certain enchanted forest" (Smullyan, 2012) there is a number of talking birds. Capital letters A, B, C denote arbitrary birds, x and y stand for arbitrary variables. "Given any birds A and B, if you call out the name of B to A, then A will respond by calling out the name of some bird to you; this bird we designate by AB. Thus AB is the bird named by A upon hearing the name of B [Smullyan, 2012, p. 63]. Smullyan adds then that A's response to B does not have to be necessarily the same one as B's response to A. Therefore one cannot simply assume that $AB = BA$. Another condition that Smullyan introduces to his readers is composition of three functions. A's response to the BC is not equal to AB's response to the bird C: $A(BC) \neq (AB)C$.

For each puzzle a different set of pre-assumed conditions can be utilized. As a result, we cannot form an universal axiomatic-like system for every puzzle. This is why ND seems like a proper system for this kind of task.

To solve some of the first puzzles, two conditions are introduced. First one is described by presuming that some bird's x response to itself is x. As a result first condition expresses a response of a bird M to bird x. This bird M is denoted by term of the form:

$$Mx = xx$$

This way Smullyan provides an easy introduction to one of the well-known combinators. Combinator M is called by Smullyan *Mockingbird*. Indeed, the result of applying it to a given argument is its duplication. This combinator will be used as one of the preconceived conditions—it will label a leaf in a tree.

Second condition expresses a property of composition. For any three birds: A, B, C (which could be non-distinct) we can say that the bird C composes A with bird B if for every bird x the condition expressed below holds:

$$Cx = A(Bx)$$

As the third condition, Smullyan uses the following function:

$$Ax = Bx$$

Smullyan describes it in a natural language as follows: if the bird's A response to the bird x is the same as the response of B to the latter one, we can say that A and B agree with each other on the bird x. What needs to be noticed is the fact that this expression is a variation of a *Mockingbird* condition. Two of the puzzles will show that—by being solved in two way (using each of these conditions).

Two last conditions used in few puzzles are two of the following ones: *Kestrel* condition:

$$(Kx)y = x$$

and the *Lark* condition:

$$(Lx)y = x(yy).$$

Additionally, if we would explain that for any bird x the following condition holds $Ax = B$, we can say that A is fixed on B. The same thing can be said about the *Kestrel* condition—bird Kx is fixed on bird y.

3.2 Syntax and rules

Rules that can be used to reach a solution for a given puzzle are simply an adaptation of the deduction rules introduced in the Combinatory logic section. For different puzzles a different set of pre-assumed conditions can be considered. Therefore we have to omit referring to the said conditions as universal axioms (however, when presenting a solution of a puzzle we will refer to them as *conditions-axioms*). They do not hold universally, we can only assume that for a specific scenario they are valid. As a result we can assume that both upper and lower case letters in these pre-assumed conditions can be easily substituted for other terms, (for instance in the composition condition), unless it is stated otherwise (in the case of *Mockingbird*, *Kestrel* or *Lark* condition). The set of rules is the following (A, B, C and x are denoting arbitrary terms of CL language):

$$(1)\ \frac{A=B \quad B=C}{A=C} \qquad (2)\ \frac{B=A}{A=B}$$

$$(3)\ \frac{B=C}{AB=AC} \qquad (4)\ \frac{B=C}{BA=CA}$$

$$(5)\ \frac{(B=C)[x/A]}{B[x/A]=C[x/A]}$$

Symbol "=" can be also understood as a case of reduction relation—a process of computation, where functions are applied to expressions (Kluge, 2006). This way, primitive combinators in the puzzles can behave like functions. Rule (1) is an example of transitivity. If a given function A can be reduced (or is equivalent) to function B and additionally function B can be reduced to some function C, we can assume that function A is equivalent to C. Rule (2) expresses commutative property. We can easily switch sides of terms as there was a deduction rule expressing the commutativity property. Rule (5) shows minimalistic approach to the notion of substitution. As there are no bound variables in the CL expressions, no additional restrictions have to be added. The process of substitution is described in Definition 4. Almost any term in a given condition can be substituted for any other function, besides terms M, K and L. The set of rules is sufficient to solve a numerous Smullyan's puzzles, however in this paper we will present few simple examples of its' use, as an introduction to the system. Additionally, a set of pre-assumed conditions can be expanded by functions that have been proved by applying the above rules. As a result, we can control the excessive expansion of the derivations (trees).

A solution of a given puzzle is presented in the form of a finite, labelled tree (Troelstra and Schwichtenberg, 2000). By tree we understand a partially ordered set (X, R) of nodes. Each of these sets is linearly ordered, with the lowest element $x \in X$ called the root of the tree. Maximal linearly ordered subsets of X are called branches. If a branch is finite it ends in leaf. If, moreover, for a given elements x, y we have xRy, then y is called a descendant of x. Additionally, if there is no node between x and y, then y is called an immediate descendant of x. By adding the following function to the tree: $n : N \to L$, where N is the set of nodes of the tree and L is the set of labels, we obtain a labelled tree.

Definition 3.1. *A resolved puzzle is a finite binary tree labelled with combinators which meets the following conditions:*

(i) the root of the tree is labelled with a combinator to be proven;
(ii) the leaves of the tree are labelled with pre-assumed conditions for a given puzzle;
(iii) for any other internal node that is not a leaf:

- *if the node has one immediate descendant—there exists a rule such that the immediate descendant is labelled with the premise of the rule and the node is labelled with the conclusion of the rule,*
- *if the node has two immediate descendants—there exists a rule such that the descendants are labelled with the premises of the rule and the node is labelled with the conclusion of the rule.*

3.3 Examples

Puzzle 1.

In the first puzzle we are ought to answer the question: can we say that every bird in the forest is fond of some other bird? To denote that bird A likes bird B, the following condition is introduced: $AB = B$. In order to check if this rumour is true, a derivation can be built:

$$\cfrac{\cfrac{Mx=xx}{MC=CC}\,(3,4) \qquad \cfrac{\cfrac{\cfrac{A(Mx)=xx}{A(MC)=CC}\,(3,4)}{CC=A(MC)}\,(2)}{}\,(1)}{\cfrac{A(MC)=MC}{AB=B}\,(3,4)}$$

We start building tree by labelling its leaves with conditions-axioms. The *composition condition* is used in its modified version. We know that for any three birds A, M, x the condition holds, therefore it can be expressed as bird x composing bird A with *Mockingbird* responding to x. As the *Mockingbird* condition holds for every bird in the forest, x can be substituted with C. The substitution rules are used in one step on both sides of the expression. In order to apply transitivity rule, the commutative property has to be utilized. This derivation could end on combinator $A(CC) = CC$, as it expressed A fondness of bird CC, but it can be reduced to $AB = B$.

Puzzle 2.

In the second puzzle we have to check if there is *an egocentric bird* in the forest. *Egocentric bird* (that can be also called *a narcissistic bird*) is one that is fond of itself, e.g. $CC = C$.

$$\cfrac{\cfrac{MC=CC}{CC=MC}\,(2) \qquad MC=C}{CC=C}\,(1)$$

As we proved in the first derivation that the condition of fondness is applicable in the forest, we can use it as an pre-assumed condition. Therefore we can add $AB = B$ to the set of conditions-axioms. This condition is applicable to the *Mockingbird* as well—it can be fond of a given bird C. We can also use the standard *Mockingbird* condition, but substitute C for x: $MC = CC$. Similarly as in the first puzzle, before applying transitivity rule, a commutative rule needs to be used. Because we can add to the set of conditions ones that have already been proven, like $AB = B$, the obtained tree is not excessively expanding.

Puzzle 3.

For the next two puzzles, a *Mockingbird* condition is replaced with the *two agreeable birds* condition.

$$\frac{\frac{Ay=Hy \quad Hy=x(Ay)}{Ay=x(Ay)}\,(1)}{x(Ay)=Ay}\,(2)$$

In the above derivation we proved the *fondness* condition using the *agreeable bird* condition. The latter one is a variation of the *Mockingbird* condition—after all *Mockingbird* is agreeable, therefore the derivation has a similar form to the one presented before. We know that for some bird y two given birds A and H agree on. Composition condition holds for any three birds, hence H can compose x with Ay. Two conditions that label the leaves have to the forms $A = B$ and $B = C$, so we can derive $A = C$. Two obtained terms in this combinator can switch places and as a result we showed that $x(Ay) = Ay$.

Puzzle 4.

Conditions-axioms for the last puzzle presented in this essay are composition condition and *agreeable birds* condition:

$$\frac{\frac{\frac{Ex=Cx \quad Cx=A(Bx)}{Ex=A(Bx)}\,(1)}{A(Bx)=Ex}\,(2) \quad Ex=D(Bx)}{A(Bx)=D(Bx)}\,(1)$$

We are to prove that if some bird A is *agreeable*, then there exists bird D that is *agreeable*, too. In order to do that, we modify the conditions to express that two given birds E and C agree on bird x and bird C is one composing A with Bx. Transitivity rule can be utilized combined with the commutative one. A pre-assumed condition can be added to the tree at any point. Thus we can assume that bird E composes D with Bx. The obtained schema is now the same as in the transitivity rule—as a result we showed that if A is *agreeable*, then so is D.

Puzzle 5.

This puzzle is a variation of the second puzzle. Assuming that composition condition and *Mockingbird* condition of the first puzzle hold and given the existence of a kestrel K, we have to prove that at least one bird in the forest is *hopelessly egocentric*.

Natural Deduction Method for Solving CL-based Puzzles

$$\cfrac{(5)\ \cfrac{Mx=xx}{MC=CC}\ \ \cfrac{\cfrac{K(Mx)=Cx}{K(MC)=CC}\ (5)}{CC=K(MC)}\ (2)}{(1)\ \cfrac{\cfrac{K(MC)=MC}{(5)}\ \cfrac{K(MC)=MC}{KA=A}}{(KA)x=Ax}\ \ \cfrac{(KA)x=A}{A=(KA)x}\ (2)}{Ax=A}\ (1)$$

We know that there is the *Mockingbird* in the forest. Suppose that there is a bird C that is being mocked. We can therefore substitute C for x. We also know that the composition condition holds for four given birds in the forest. We can assume that bird C composes *Kestrel* with the *Mockingbird*. Now we can utilize the transitivity rule and as a result obtain $K(MC) = CC$, which can be transformed into $CC = K(MC)$. We also know that there is the *Kestrel* in the forest. Suppose bird KA is fixed on bird A. Again, we can use the commutativity property on this function. The only step left to prove that there is *a hopelessly egocentric bird* in the forest is the transitivity rule.

Puzzle 6.

The last puzzle is the most complex one in terms of the size of the tree. However, it has been based only on one of the pre-assumed conditions. Again, we are seeking to prove the same condition as in the second puzzle—*an egocentric bird*. In this case Smullyan shows that the only information needed to prove existence of *an egocentric bird* is that the forest contains the *Lark L*.

$$\cfrac{(5)\ \cfrac{(Lx)y=x(yy)}{(LL)y=L(yy)}\ \ \cfrac{(5)\ \cfrac{(Lx)y=x(yy)}{(Lx)(Lx)=x((Lx)(Lx))}\ \ \cfrac{x((Lx)(Lx))=(Lx)(Lx)}{LL(y)=y}\ (5)}{(4)\ \cfrac{L(yy)=y}{(L(yy))y=yy}}\ \ \cfrac{(5)\ \cfrac{(Lx)y=x(yy)}{(L(yy))y=(yy)(yy)}}{(yy)(yy)=(L(yy))}\ (2)}{(yy)(yy)=yy}\ (2)$$

As in this case the only pre-assumed condition is the *Lark* condition, the essential rule is the substitution. Starting with the left branch of the tree—we begin with substituting constant L for the x. On the second branch we also have to start with the substitution. The obtained function is $(Lx)(Lx) = x((Lx)(Lx))$ which is a result of substituting term Lx for y. After using the commutativity property, we can distinguish two terms appearing in the resulting function—x and $((Lx)(Lx))$. We can substitute LL for the first one and y for the latter one. Now we have two nodes labelled with combinators that can be merged with respect to the transitivity rule. On the last branch the substitution process starts with replacing x with yy. In order to obtain *egocentric bird*, we have to utilize sequentially commutativity, transitivity and again commutativity rule.

4 Final remarks

As shown above, adapting ND method for solving CL-based puzzles results in a clear set of instructions on how to proceed with the search for an answer. However, we are yet to specify how certain operations should be executed in order to obtain a minimal tree (the simplest solution). Even Smullyan stressed (in regards to the Lark puzzle) that for some of the puzzles there could exist a different and shorter solution. For the algorithmic purposes it seems that the simplest process of building tree could start from the root labelled with a (to be proven) combinator and then by working our way up to the leaves labelled with pre-assumed conditions. Nonetheless, in this case it seems that the one crucial rule that could affect the tree growth to a great extent is transitivity. Introducing two new terms should be in some way governed by a set of restrictions; it should not by in any way random. Therefore, the next step in improving this method would be analysing different solutions in order to obtain and specify such restrictions. Implementing the above method in programming language could be one way of achieving that.

Overall, Smullyan's CL-based puzzle provide a simple introduction to the topic of function-based systems. As a result, Smullyan's descriptive approach could possibly encourage students to learn more about formal sciences and to have less anxiety related to such subjects. Besides the educational purposes for which they can be beneficial, CL-based puzzles give us an insight into meta-theoretical properties of CL system and let us develop new ways of utilizing them.

References

Aszalos L (2002) Automated Puzzle Solving. Journal of Applied Non-Classical Logics 12:99–116

Barendregt H (2013) The Lambda Calculus: Its Syntax and Semantics. Studies in Logic and the Foundations of Mathematics, Elsevier Science

Casimir R (1987) Prolog Puzzles. SIGPLAN Notices 22:33–37

Hindley JR, Cardone F (2006) History of Lambda-calculus and Combinatory Logic. Swansea University Mathematics Department Research ReportNo MRRS-05-06

Jerardi W (1987) Puzzles, Prolog and Logic 22:63–69

Keenan D (2014) To Dissect a Mockingbird: A Graphical Notation for the Lambda Calculus with Animated Reduction

Kluge W (2006) Abstract Computing Machines: A Lambda Calculus Perspective. Texts in Theoretical Computer Science. An EATCS Series, Springer Berlin Heidelberg

Kolany A (1996) A general method of solving smullyan's puzzles. Logic and Logical Philosophy 4, DOI 10.12775/LLP.1996.004

Ramsey BD (1986) The lion and the unicorn met PROLOG. In: SIGP

Smullyan R (2012) To Mock a Mocking Bird. Knopf Doubleday Publishing Group

Sorensen M, Urzyczyn P (2006) Lectures on the Curry-Howard Isomorphism. Elsevier

Sprenger M, Wymann-Böni M (1993) How to decide the lark. Theoretical Computer Science 110(2):419 – 432

Statman R (1989) The word problem for smullyan's lark combinator is decidable. Journal of Symbolic Computation 7(2):103 – 112

Troelstra AS, Schwichtenberg H (2000) Basic Proof Theory, 2nd edn. Cambridge Tracts in Theoretical Computer Science, Cambridge University Press, DOI 10.1017/CBO9781139168717

Wadler P (2015) Propositions As Types. Commun ACM 58(12):75–84, DOI 10.1145/2699407

Wolfengagen V (2003) Combinatory Logic in Programming./edited by dr. l. Yu Ismailova Moscow: Center JurInfoR 336

The Measurement of Factive Deductivity: a Psychological and Cerebral Review

Paula Álvarez-Merino and Carmen Requena and Francisco Salto

Abstract There are normative differences between deduction and induction, but are there factual differences between them? The state of the art offers all logically possible answers to this question. Let the Hypothesis of Factive Deduction state that one and the same deductive phenomenon occurs at normative, psychological and neural levels. A systematic review of psychometrical and neural tests was conducted in the search for measures to empirically confirm or refute the Hypothesis. The review identified 27 psychometrical and 58 neural measures of which 15 psychometrical and 16 neural tests met criteria for inclusion. Deductive properties such as: logical validity, probabilistic validity, logical vs. relational complexity, format invariance, computability, integration and formality level have been operationalized. For any given deductive component, we determine if it is measured, explicitly assessed, present or absent in each test. Results show the absence of deductive measures and the presence of inconsistent constructs corresponding to distinct notions of deduction. Non-explicitly deductive components (microdeductions) in reasoning tests are identified. The introduced deductive variables intend to contribute to a future confirmation or refutation of the Factive Deduction Hypothesis.

Key words: Deduction, Inference, Reasoning, Measure, Logical complexity, Relational complexity, Deductivity, Visual, Linguistic

---

Paula Álvarez-Merino
Universidad de León, Spain
e-mail: paulaalvarezmerino@gmail.com

Carmen Requena
Universidad de León, Spain
e-mail: c.requena@unileon.es

Francisco Salto
Universidad de León, Spain
e-mail: fsala@unileon.es

1 Introduction

Deductive inference is a possibly modest but crucial component of scientific and everyday rationality. There are clear normative differences between deductions (Hintikka, 1999; Johnson-Laird and Khemlani, 2013; Koralus and Mascarenhas, 2013) and other forms of inference, such as inductions (Johnson, 2017) or abductions (Magnani, 2011). Deductive inferences are *justified* if they preserve properties of their premises, notably their truth, or alternatively their probability, their credibility and their demonstrability, among other alternative metaproperties, while inductive inferences are not *justified* because of the properties they preserve, but rather for maximizing the information and utility of their premises. Deduction is *justified* by means of proof while induction is usually not. Deductive inference (if classically valid) necessarily *implies* its conclusions, while inductive inference does not. Therefore, there are clear normative or argumentative differences between deductive and non-deductive inferences.

However, it is far from clear if there are any *factual*[1] differences between deductive and non-deductive cognitive inferential events or processes. Even if the science of deduction is as old as science itself and has traditionally mingled psychological deduction processes with normative deductive procedures for justification, demonstration and proof, we must distinguish between deduction as a *cognitive* and as an *argumentative* instrument both in science and ordinary reasoning (Harman, 1984). As cognitive inferential events, deductions are psychological or neural processes, while taken as propositional arguments, deductions are instances of implications between propositions (Harman, 2002), or Bayesian probabilistic relations (Oaksford and Chater, 2007). In the face of the question: "are there really factive differences between the psychological processes of deduction and induction?" the recent literature offers all logically possible answers:

1. Positive answers. One line of work, the most extended and akin to the traditional comprehension of deduction, acknowledges *a priori* differences between the processes of deducing and inducing (Dogramaci, 2017; Evans and Over, 2013; Evans

---

[1] A process is *normative* when it happens only if it should happen, that is, if it instantiates certain values or certain abstract properties or rules, as opposed to *factive* or *factual* processes which take place precisely when they just happen. Sunrise is a factive process because there is no bad or good sunrising, while digestion, argument or excuse are normative processes. The precise nature of the distinction is philosophically disputed since any factive process could be taken under some arbitrary norm. Even in the absence of a fully transparent notion, in the fields of ethical, semantical and logical valuation there seems to be a non-arbitrary distinction between normative and factive processes. Factive processes are realized and described in terms of facts, while normative processes are justified in terms of norm satisfaction. Certainly, the frontier between both types of processes is not always sharp and we do not need to presuppose that it is a not crossable frontier. Even though it is usual to present reasoning as the set of psychological processes producing inferences (Johnson-Laird, 2008), these inferences, either practical or theoretical, are not adventitious events. Only if their conclusions are pertinent, repeatable and non arbitrary can inferences count as reasonings. Therefore, reasoning is a normative process and for this reason it is subject to criticism, rebuttal and learning as opposed to mere contradiction or repetition. Traditionally, on the base of different forms of justification or legitimation of their conclusions, two basic forms or normativity have been distinguished in reasoning: deductive and inductive (and on occasions also abductive).

et al, 2015; Schechter, 2013; Wilhelm, 2005; Williamson, 2003). It is remarkable that this perspective is compatible with rival conceptualizations of deduction and of normativity in general, as shown by the references just given. In particular, both the counter-example conception of deduction (Schaeken, 2007; Schroyens et al, 2003) as also the calculative conception (O'Brien and Manfrinati, 2010) both in their truth functional (Johnson-Laird, 2008) and probabilistic (Johnson, 2017) versions assume that there are a priori differences between deductive and non-deductive inferences. Specifically, Evans and Over (2013) offer *a posteriori* reasons to qualify deductions to be as necessary as truth or probability validity preservators.
2. Gradual answers. Another line of research records gradual factive differences between deducing and inducing (Heit and Rotello, 2010). Deductive inferences are sensible to properties such as logical or probabilistical validity in contrast with properties such as longitude or number of premises (Heit, 2015). A second *a posteriori* approach to deductivity is offered by Rips (Rips, 2001; Singmann and Klauer, 2011; Singmann et al, 2016) offering partial and indirect empirical support for factive deductivity in linguistic propositional inferences.
3. Negative answers. A third approach assumes that there are no factive differences between mental or cerebral inductive and deductive processes, as explicit in the words of Oaksford (2015): "... tasks are not deductive in and of themselves. What function a task engages is determined by the empirically most adequate computational level theory of that task". From another perspective, a different negative understanding of gradual data is offered in (Stephens et al, 2018).

This state of the art is extraordinary and shows a clear incongruency between the normative features of deduction (which are manifestly distinct from inductive ones) and the factive features of deductive processes (which are not manifestly so). In this regard, we cannot depart from the existence of *factual* features of deductions as being opposed both to non-deductive inferences (inductions, abductions) and to non-normative deductivity criteria[2]. This paper reviews the state of the art regarding factive (psychological or neural) measures of deductive inference. Instead of presupposing that deduction is one and the same phenomenon present at the three distinct levels of analysis proposed by Mart (1982) and Baggio et al (2015): normative, psychological and neural levels, we search for measures to confirm or refute deduction as a factive hypothesis.

2 Antecedents: psychometry of deduction

Early studies on reasoning measurement focused on the psychological assessment of the factive distinction between inductive and deductive reasoning. These studies

---

[2] This normative/factive tension is certainly not exclusive of deductive phenomena and moreover it is deeply rooted in historically blurred relations. Arguments for the two opposing views are found in Husserl, 1900 and Woods, 2016.

resulted in disparate results by Colberg et al (1985) and Shye (1988) which were finally systematized by Carroll et al (1993). Four basic dimensions of reasoning emerged:

- operational, which is mainly the deductive/inductive distinction
- content, which includes sentential, figural and visual types of representation involved in reasoning
- instantiation, which distinguishes between abstract items (such as pseudowords) and concrete items
- strategies, which discriminates among different ways of tackling the same inferential problem.

With these measures, the work of Wilhelm (Schroeders and Wilhelm, 2010; Wilhelm, 2005) confirmed that inductive and deductive procedures equivalently contributed to those four basic factors in terms of reasoning measurement. Wilhelm concluded that deductivity versus inductivity has less explanatory impact on reasoning than the type of representations involved (content). In subsequent studies deductivity was not a relevant variable in psychometrical studies on intelligence and was substituted with non-normative variables such as working memory (Dehn, 2017; Kyllonen and Christal, 1990) or even executive function (Alvarez and Emory, 2006; Morra et al, 2018). Beyond the first generation of reasoning measures, Stanovich (Shynkaruk and Thompson, 2006) conceived enriched algorithmic and analytic assessment tools for non-normative deductive reasoning (Markovits et al, 2017; Prowse Turner and Thompson, 2009; Shynkaruk and Thompson, 2006; Thompson et al, 2011), which preserves factive properties such as subjective confidence (as opposed to logic or probabilistic confidence) and feeling of rightness. Notice that "deductive" in this context lacks any normative status: an adventitious conclusion is deduced from a set of premises if it is triggered from them. Sometimes the conclusion is to be qualified as "new", "informative" or "coherent" (del Carmen Crivello et al, 2016), but the definition of those notions seems harder than deduction itself. Wilhelm's indifference result is only based in general reasoning measures and does not include specifically deductive properties. In order to verify the factive irrelevance of deductivity, abstract properties defining deduction need to be considered. These properties are not restricted to any particular kind of reasoner (biological or mechanical), nor restricted to any specific sort of support (visual or linguistic), since they are instantiated by humans, animals, machines, groups, abstract objects and institutions. We propose and schematically present the following general properties, which have been previously validated (Alvarez Merino et al, 2018) and will be operationalized as measure variables later in the paper. Validity.

At an abstract level of analysis, inference patterns can be considered as logical consequence relations (in the abstract sense of Wójcicki (2013) and its generalizations) preserving desirable properties, such as functional bivalent truth (Gödel, 1930), computability (Church, 1940), constructive demonstrability (Martin-Löf, 1975), probability (Adams, 1996) or game-winning strategy (Hintikka, 2002). It won't be described here how "preserve" is defined in each case, since it is a well-documented technical issue. According to this conception, the principles of implication are normative because they warrant the preservation of semantical or epistemical properties at the metalogical level. Therefore, validity is not a factive property of psychological or neural events, but it is a precise and accessible property that eventually co-variates with factive properties.

There are several inequivalent (often competing) normative properties preserved by deductive processes. Significant examples include classical truth functional bivalent validity, probabilistic p-validity, modal or index-relative validity concepts, constructive validities, substructural and relevant validities. Some properties preserved by conclusions of deductions are not normative, such as the statistical frequency of the conclusion, or its subjective acceptance or "feeling of rightness" (Markovits et al, 2017). Notice that validity is associated with a way of justifying any property instead of with any particular property itself. In this sense, validity is a metaproperty of inferences, or more precisely of the arguments justifying valid inferences. Logical validity is an extremely abstract and demanding property, and reasoners often engage in less abstract inferences in which normative properties are preserved at a minor scale. For example, a mathematical deduction can be constructive or not involving distinct valid criteria, while deducing the translation of a word involves less abstract normative criteria. In this sense, (meta)logical pluralism is assumed in this proposal.

2.1 Computability

A process being computable or recursive implies that it is deductive only in the abstract sense of being normatively definable by deductive rules or axioms. This global equivalence does not imply that any computable factive process is deductive (for example, pattern recognition processes). Moreover, most factive processes lack explicit negation and will at most semi-compute (Boolos et al, 2002) their conclusions. However, the logical process of deduction does not extensionally coincide by chance with the recursive process of computation (Kennedy, 2017). This coincidence is based on the fact that distinct and independent descriptions of recursive functions happen to be all globally equivalent, an invariance that Gödel called "the miracle of recursion" (Gödel, 1995). An eventually distinctive factor of deductive inference is based on this coincidence.

Deductive processes seem to present themselves in systematizable, automatizable and recursive ways which contrast with some non-deductive processes, at least at an abstract level. Factors such as monotonicity, compositionality and repeatability seem peculiar of many deductive procedures and remain often absent in inductive inferences. A significant but scarce number of experimental results show some factive vestiges of compositionality at the neural level of deductive reasoning, as manifested in the work of Reverberi and Baggio (Baggio et al, 2015; Reverberi et al, 2012).

2.2 Logical/relational complexity ratio

The key idea into refuting or confirming factive deductivity is to identify and measure the eventual presence of differences between the relative impact of logical and non-logical complexity in deductive and inductive inferences. The idea is found from a psychological perspective already in Rips' work (1994; 2001) and from a neural perspective in (Baggio et al, 2016) with the strategy measuring the neural impact of increasing logical complexity while sentential complexity is fixed. Research has also applied Signal Detection Theory (SDT) in truth functional (Rotello and Heit, 2009) and probabilistic (Lassiter and Goodman, 2015) settings to measure the relative impact of validity over factive reasoning variables such as length.

Logical complexity is usually defined in sentential contexts in terms of the number of occurrences of logical constants. Another form of complexity concerns the factual inferential process, defined in terms of time, space, energy and syntactical properties such as length, lexical variety, grammatical clauses or relational complexity (Cocchi et al, 2013). This processing complexity concerns the number of variables or items present in the deductive inference (which is factively severely restricted in the human brain), the number and variety of lexical units and the extension of inferences and arguments. In deductive inferences, the relation between both forms of complexity is different from the relation found between the same complexities in inductive inferences. As a result, a putative factive measure of deductivity emerges. Measurement of logical complexity in plastic or visual contexts is largely an open question (Alemany, 1998; Shimojima, 2015), even if all kind of inferential formats allow a distinction to be made between low logical complexity (associated with Modus Ponens inferences) and higher logical complexity in the presence of negation or other modalities (*Modus Tollens*).

2.3 Format invariance

Sentences, images and actions are three dramatically different formats for eventually the same deductive events. Most factive properties are obviously not format invariant, and therefore their presence is a pertinent property for our purposes. Moreover, both psychological and neural factive research pose distinct methodological and temporal experimental designs for these formats. We shall distinguish three kinds of format: linguistic (usually sentential), visual (or more in general, plastic) and agentual (agency-based reasoning). Existing psychometrical measures for reasoning include both linguistic and visual formats.

One of the main results in the neural study of deductive inference is the heterogeneous character of its brain networks (Monti et al, 2012b; Morra et al, 2018) which includes visual circuits, latencies and sources distinct from linguistic cerebral connectivities. How inferences with the same logical structure but different format are processed in the brain is an open issue.

2.4 Formality levels

Tokens, words, pseudo-words and formal signs are increasingly abstract instances of deductive tasks in tests. Wilhelm also distinguished formality levels in reasoning tasks and it is a common strategy in neurocognitive research (Hagoort, 2013). A semantical property such as truth is preserved in a particular model or preferred interpretation. We may assume that this is a 0-level of formality and coincides, in the case of this property, with extensional truth in such model or interpretation. At a higher level of formality, we may consider the preservation of properties by definition through several intended models. At this 1-level of formality we locate truth-in-classes-of-models, corresponding to definitions and alleged analytical truths. In this coarse analysis we find the purely formal logical inferences at a higher level of formality (2-level).

2.5 Integration

Deducing does not seem to concern adventitious items but rather conclusions that are non-arbitrarily integrated with their premises. The psychology of reasoning does not offer us a precise already established notion of premise integration, but we can employ more modest conceptual tools from the logical literature, where entailment (Routley and Meyer, 1972), relevance (Dunn, 2015) and variable sharing (Méndez et al, 2015) offer partial surrogates of integration. Moreover, the measure of integration should be viable both for arguments and for plastic inferences (see how this property is operationalized below).

If deduction is not only a way of justifying inferences but also a factive process taking place at computational, psychological and neural levels, then psychometrical and brain measures should offer hints of it. The objective of this work is to review the psychometrical and neural measures of deductivity, particularly to verify the extent, consistency and accuracy with which deductive properties are present and measured in factive contexts. Operationalized versions of deductive properties are defined and contrasted with psychometric and neural tests. The point of the whole work is to set up the experimental bases to verify or refute the hypothesis of factive deduction.

3 Methods

A comprehensive review was conducted of psychometric tests and cerebral measures applied on reasoning tasks which describe themselves as deductive. The review is focused specifically in measures and tests that are explicitly directed to assess deductive tasks irrespective of their nature. The review includes two distinct comprehensive reviews. One of them surveys validated tests and subtests on deductive reasoning, while the second one examines ad hoc deductive measures which are not validated tests but rather measures designed for specific brain research purposes, usually in the field of brain correlates of deductive reasoning. Both reviews systematically cover all pertinent contributions published or in press prior to October 2018.

The psychological tests review includes validated tests explicitly targeting deductive constructs including deductive procedures embedded in other inferential processes. Only academic tests have been included, while commercial or business-oriented tests have been excluded. Tests were directly searched for in tests repositories: APA Psyctests, EFPA, Academic Catalogs, Research Institution Repositories and research institutions. Other tests have been indirectly searched for via publications in computerized databases (PsychInfo, PsychLit, PubMed, CogPrints: Cognitive Sciences Eprint Archives, CiteSeerX, dblp computer science bibliography, MathSciNet, Philosophy Documentation Center eCollection, ScienceOpen). Non-academic repositories with involvement in commercial activities have been excluded from the study. The search included three components corresponding to the following three sets of terms: "reason*", "inferen*", "deducti*", "logic*", "formal", "probabili*", and "test", "subtest", "measure", "evaluation", "assessment". For example, a typical search is the following Boolean combination of terms: "DEDUCTI*" AND ("REASON*" OR "INFEREN*") AND ("TEST" OR "MEASURE" OR "EVUALUATION"). Particular search engines may require simpler descriptions of logically equivalent searches, like such as DEDUCTI* AND REASON* AND TEST. Beyond the direct search for measures and tests in tests collections, additional indirect searches of tests were conducted surveying computerized databases (PsychInfo, PsychLit, PubMed), published reviews and meta-analyses (Khemlani and Johnson-Laird, 2012; Rips, 2001) and performed hand searches of journals (e.g., Reasoning; Applied Cognitive Psychology; Cognition; Cognitive Technology; Psychological and Educational Measure), and book chapters (Evans and Over, 2013; Oaksford and Chater, 2007; Stenning and Van Lambalgen, 2012). Additional studies were identified from reference lists in retrieved articles. For example, the test TOLT (Tobin and Capie, 1981) was found in the second indirect search for tests and not in the first one.

Inclusion criteria for tests in direct and indirect searches were based on the psychometrical validation of the measure and the presence of key terms in their description or in their constructs. 27 Tests complying with Psyctests APA or EFPA criteria for identification of evidence-based measures were scrutinized and excluded if they employ deductive abilities as a measure of other clinical, educational, social, cognitive or emotional phenomena. For example, the TOGLIA test was excluded for this reason. Only tests in English and Spanish have been included in the review. Adaptations and translations of deductive tests are not considered as new measures. As a criteria of test identity, it has been assumed that any subtest satisfying the inclusion criteria will be considered a test (for example, subtests such as "Figure Weights" of WAIS-IV are considered as distinct tests in this review) (Wechsler, 2008).

For the sake of completeness in the review of factive measures, cerebral and neural measures associated with deductive inferences have been also reviewed. Some significant methodological difficulties are faced, since the review of cerebral results on deduction concerns *ad hoc* measures of deductive reasoning and inference. No specific test has been found in APA or EFPA, therefore the review focused on published results and not directly on measures.

The psychological tests included in the review are the following: BPR (Batería de Pruebas de Razonamiento, Elosúa and Almeida 2016), WAIS IV FIGURE WEIGHTS, ARITHMETIC, MATRICES, CUBES (Wechsler, 2008), SX scales, FX scales, PROPOSITION TEST (Schroeders and Wilhelm, 2010), EROTETIC (Koralus and Mascarenhas, 2013), R30, RP30 (Seisdedos, 2002), TOLT (Tobin and Capie, 1981), DFT (DEDUCTIVE FLEXIBILITY TEST Żyluk et al 2018).

The search added a fourth component: "Neur*", "Brain", "Psychol*", "Cognitive" to the previous three components and the same computerized databases were employed. 58 eligible results were individually filtered down to 16 pertinent papers. Metanalysis and reviews such as Prado et al (2011) and Goel et al (2017) were used but not included in the review. Since the limited size and number of papers allows it, an individual review of authors and journals resulted in 29 final pertinent tasks included in 16 works.

3.1 Variable operationalization

Deductions have specific normative properties. In order to investigate if there are also factive properties peculiar of deductive inferences, correlations are between measured variables in the reviewed tests and specifically deductive variables identified from the literature are searched for: validity, computability, format invariance, logical/relational complexity, formality level and integration. Some of these variables need to be operationalized in order to verify their presence/absence in tests.

Validity is operationalized in this review with two specific measures: classical logical validity (bivalent and truth-functional) and p-validity or probabilistic validity as defined by Adams (Adams, 1996).

Computability can be operationalized with several measures. In this review we have selected the presence/absence of semi-computable procedures mainly based on the presence/absence of full classical negation. Visual inferential tasks lack mechanisms to codify negation and play an important role in matrix reasoning. Compositionality and monotonicity have not been included as deductive features.

Format invariance: it is operationalized by means of specific examples of instances of the same logical scheme in distinct tasks with different visual and linguistic format. For example, disjunctive syllogism can be presented in visual (Dove, 2016), agentual (Dogramaci, 2017) or linguistic formats.

Complexity: it is operationalized with the long-term reaction of an inference pattern to increases in logical complexity in contrast with increases in relational complexity. If we take pure truth functional complexity as the length of the shortest logically equivalent propositional formula, we obtain an easily definable magnitude for sentential logical complexity. This direct measure (both if it is understood in Boolean or in non-classical terms) is severely limited as a factive measure, since the psychological process of deducing is not isomorphic to the logical process obtaining the same conclusions from the same premises. Among many other examples, mental models research on deduction (Johnson-Laird, 2008) has found that increasing logical complexity does not correlate positively with the duration of the processing nor with the presence of biases. Therefore, direct complexity measures are not useful to assess deductivity, but rather indirect measures formed by tuples ⟨direct logical complexity value, processing complexity value⟩. At the factive neural level, changes in wave amplitude and brain connectivity associated with changes in logical complexity have been diversely interpreted. Monti et al (2007, 2012a) contemplate neural processes directly dependent on deductive complexity, while Prado (Prado et al, 2011; Houdé and Borst, 2015) interpret the neural correlates of deduction as dependent on semantical processing.

Integration: it is operationalized in terms of explicit relevance (variable sharing property, VSP) (Méndez et al, 2012). A premise is expressly integrated or relevant to a conclusion if it literally shares one of its variables (propositional, lexical or plastic). In consequence, a conclusion is integrable just in case it literally contains contents of its premises. For example, of the premises: All big and central apartments are expensive. This central apartment is not small we consider integrable the conclusion: "This apartment is expensive". Neither the following true conclusion: "Madrid is the capital of Spain" nor the valid conclusion: "If all are happy, someone is happy" are. Logically or probabilistically valid deductions do not require integrable premises, and of course inversely integrable inferences are not in valid in general. Integration is a significant variable in deductive reasoning (Evans, 2002; Skovgaard-Olsen et al, 2016), and its statistical significance overwhelms factors such as age or even validity (Merino et al, 2016).

The definition is also applied to reasoning in a plastic or visual format. A plastic inference is *integrated* if the set of premise images provides the conclusion without using auxiliary premises or choosing among alternatives. Thus, the following tasks extracted from a Reasoning Assessment tool serve as examples of integrated and non-integrated plastic inferences, respectively.

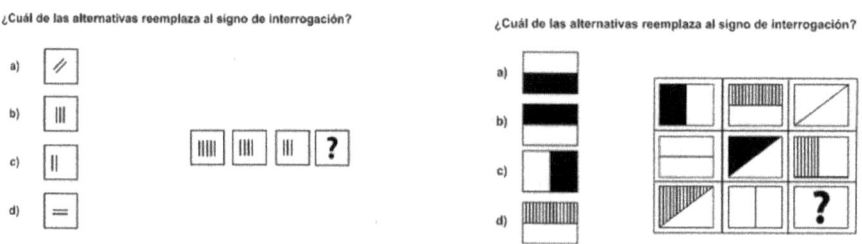

Fig. 1: Integrated inference (left), non-integrated inference (right). In both cases it is asked: Which of the alternatives replaces the question mark? In the first exercise, the conclusion appears literally, while not in the second. Explicit alternatives are not required in integrable inferences. In the second exercise, reasoning on offered alternatives is required

3.2 Results

Based on the stated inclusion and exclusion criteria for psychological tests, 15 of the 32 initially included tests were validated deductive reasoning tests involving linguistic or visual formats and therefore eligible for inclusion in the present review. No test involved an agentive format. Common reasons for exclusion involved designs that included unvalidated tests, inferential tests which lacked explicit deductive tasks, or inferential tasks with undifferentiable deductive components. 9 of the 15 published and validated tests on deductive inference are in fact subtests incorporated in broader reasoning measures and were in fact not specifically designed to measure deductive inference.

With the partial exception of validity, the reviewed tests do not in general measure the deductivity properties identified in the literature. Even if most properties are in fact present in deductive tasks, they are not measured. All tests exhibit at least one format and include distinct levels of logical and relational complexity, but format invariance and logical/relational ratios are not present in the psychometrical measures. Logical validity is measured in 7 of the 15 tests, while probabilistic validity is absent. Unintegrated inferences are avoided by all reviewed tests. Six tests exhibit verbal format and two of them have tasks in both formats (Elosúa and Almeida, 2016; Schroeders and Wilhelm, 2010). Most tasks are compositional and monotonic, while six tests are semicomputable and nine are computable or recursive. Only 4 tests include pseudowords or formal symbols, while most tasks and tests present high formality levels (Alvarez Merino et al, 2018).

The psychometrical results are summarized in table 1, where four influential tests which do not measure logical validity still include it as a factor of inferential success. Abstract, Matrix, Cubes, and Balance tasks (Elosúa and Almeida, 2016; Wechsler, 2008) are pattern instantiation problems which involve logical validity in an indirect way, since implicit automatic inferences must be present both in the positive and in the negative assessment of any case. We introduce the notion of microdeduction to identify this implicit deductive element, which is in fact not measured.

The cerebral review includes 16 research studies of neural activity associated with explicitly deductive tasks during the period 2011-2018, after the publication of Prado's meta-analysis (Prado et al, 2011) which systemized relational, categorical and propositional reasoning with distinct viso-spatial and linguistic circuits. Correspondingly, the studies focused on relational and transitive reasoning (Mathieu et al, 2015; Prado et al, 2012, 2010; Ragni and Knauff, 2013; Wendelken, 2015), categorical and syllogistic (Houdé and Borst, 2014; Tsujii and Watanabe, 2010), propositional (Baggio et al, 2016; Coetzee and Monti, 2018; del Carmen Crivello et al, 2016; Dunn, 2015; Liang et al, 2014; Monti et al, 2012a; Ragni and Knauff, 2013; Reverberi et al, 2012). Neither research on negation and contradiction nor theoretical or divulgative works have been included in this review. The reviewed scanning techniques are fMRI, MEG and EEG, TMS and lesion symptom mapping.

There is no reviewed neural evidences verifying of refuting the hypothesis of factive deduction. Regarding format invariance, it has been confirmed that different forms of deductive inference evoke distinct brain areas, connectivities and latencies (Liang et al,

2014). However, it has not been confirmed that visual or linguistic format determines the neural processing of deduction. Baggio et al (2016) regarding propositional connectives identifies a first superficial processing which seems to be format dependent, while a second fully logical processing could be format independent. Regarding the neural correlates of deductive inference there are two distinct sensibilities in the literature. According to Prado (Goel et al, 2017; Prado et al, 2010), the neural processing of deducing is assimilated in contentful linguistic or spatial representations. On the other hand, according to Monti and Osherson (2012), deduction evokes specifically formal processes which are dissociated from contentful language. Monti's main argument concerns the cerebral repercussions of increasing logical complexity while relational complexity remains fixed. The literature offers divergent interpretations of the scanning results. On the other hand, some basic computational properties associated with deductive processes have been restrictedly confirmed in the neural processing of ordered sequences of inferences. In this regard, special attention has been given to the compositionality of the cerebral processing of semantical (Baggio et al, 2012) and deductive inferences (Reverberi et al, 2012).

3.3 Discussion

The paper reviews psychometrical and cerebral measures of deductive reasoning in the search for factive measures confirming or refuting the distinctive existence of deductions across neural, psychological and normative levels. It has not been feasible to verify or refute by experimental means the existence of measured inferential items which are (i) normatively deductive and (ii) factively characterized by any psychometrical or neural measure. Hence, the review shows no empirical evidence (nor counterevidence) of factive deduction as one and the same type of inference across linguistic, visual and agentual formats. However, a set of variables including validity, computability, logical/relational complexity, format invariance and integration is proposed and operationalized in order to achieve confirmation or refutation of factive deduction in the future.

The absence of a consistent deductive construct lies in the center of explaining the absence of deductivity measures. First, from a conceptual perspective, we have already seen that "deductive" is used to refer to (1) logically or probabilistically valid inferences preserving semantical or epistemical properties such as bivalent truth, subjective probability, demostrability, etc., and (2) less strict normative constrains (coherence, informativity, pertinence) or even empty normative constrains if an adventitious conclusion B is deduced from a set Γ of premises just in case it is factively inferred from Γ without the intervention of other premises. Sometimes (2) is presented as a version of (1) in which deduction is not certain (del Carmen Crivello et al, 2016) or no implication. Sometimes (1) is presented as a particular monotonic and truth-functional case of (2). Both identifications are far from obvious and in both cases an experimental elucidation of the relationship between both notions of deduction requires distinct constructs for (1) and (2).

The Measurement of Factive Deductivity

Table 1: Psychometrical review. Measured: offers an explicit measure; Assessed: distinguish among levels; Present: dimension involved; - Complexity levels: 0,1,2 +: yes -: no

	Logical validity	Prob. validity	Logical compl.	Relational compl.	Computability	Format invariance	Integration	Formality level	Dicto measure
BPR abstract reasoning	microded	–	Present	assessed	Semicomp.	visual	+	1	Abstract reasoning
WAIS IV matrices	microded	–	Present	assessed	Semicomp.	visual	–	2	Perceptual reasoning
SX	measured	–	Assessed	assessed	Computable	visual	+	2	Logical reasoning
FX	measured	–	Assessed	assessed	Computable	visual	+	2	Logical reasoning
PROPOSITION	measured	–	Assessed	assessed	Computable	verbal	+	2	Verbal, deductive
EROTETIC	measured	–	Assessed	assessed	Computable	verbal	–	1	Numerical reasoning
WAIS arithmetic	Present	present	–	assessed	Computable	numerical	+	1	Numerical reasoning
BPR numerical	Present	present	–	assessed	Computable	numerical	+	1	Perceptual reasoning
WAIS Figure Weights	–	present	Present	assessed	Computable	visual	–	2	Deductive reasoning
BPR analytic	measured	measured	Assessed	assessed	Computable	verbal	–	2	Perceptual reasoning
WAIS cubes	–	–	Assessed	assessed	Semicomp.	visual	+	1	Spatial reasoning
BPR spatial	microded	–	Assessed	assessed	Semicomp.	visual	+	1	Propositional Logic
R30	microded measured	–	Assessed	assessed	Computable	visual verbal	+	2	Deductive Learning
TOLT	measured	–	Assessed	assessed	Computable	visual verbal	+	2	
DEDUCTIVE FLEXIBILITY	measured	–	Assessed	assessed	Computable	verbal	+	1	

Table 2: Cerebral review

Study	Deductive Inference	Scanning Method	n	Contrasts Included
Tsuji 2011	Categorical	rTMS	24	Congruent/incongruent
Monti 2012	Propositional	fMRI	21	Algebra/natural language
Liu 2012	Proposicional	fMRI	14	Conditional Falsification
Wendelken 2012	Relational	fMRI		Visospatial/semantic
Reverberi 2012	Propositional	fMRI		Identity, Order
Reverberi 2012	Propositional	fMRI		Awareness
Ragni, 2013	Relational	TMS	29	Model selection
Prado 2013	Relational	fMRI		Transitive
Houdé 2014	Categorical	fMRI		Inhibitory control
Cutmore 2015	Propositional	ERP	16	Conditional
Mathieu 2015	Relational	fMRI	36 children	Linear ordering/set inclusion
Baggio 2016	Propositional	fMRI	35	Logical connectives
Papageorgiu 2016	Categorical	EEG	51	Valid/paradoxical
Eimontaite 2018	Propositional	Lesion Symptom Mapping	109	Disjunction
Mathieu 2018	Numerical	fMRI		Arithmetic/spatial
Coetzee 2018	Propositional	fMRI	20	Logical complexity

It is illustrative to explain the concept of deduction appealed to in WAIS IV. The CHC (Carroll et al, 1993), theory is explicitly employed in WAIS IV's manual (Carroll et al, 1993) to define the measures of matrix reasoning as reproductions of the ability of fluid reasoning (Gf) measured by the General Sequential Reasoning subtest. According to CHC's own description, this matrix test captures deduction understood as "the ability to begin with rules, premises or given conditions and develop in one or more steps a solution to a new problem". This a priori characterization of deduction illustrates how deductive inference can be developed both in plastic or visual contexts as in linguistic or propositional contexts. It is elusive because it avoids any consistent normative approach to what counts as a deductive solution to a problem. This *a priori* characterization of deduction is not consistent with normative conditions defining what counts as a solution in deductive terms as opposed to usual, normal or even arbitrary solutions. In order to verify that this normative free notion of deduction is factively adequate, it should be experimentally contrasted with deductively normative notions.

A second difficulty for a consistent psychometrical deductivity construct is the need to test deductive inference forms across linguistic, visual and agentual contexts which require distinct temporal Windows and experimental designs. The scarcity of deductive tasks reviewed in verbal contexts probably reflects the fact that the interest in reasoning is focused on fluid abilities while verbal tasks in the selected tests are restricted to inductive reasoning such as similarities recognition, analogies and comparisons. A traditional feature of deduction is its systematic nature, which is closely related to proof and demonstrative mechanisms. However, factive deduction does not seem to be systematic. For example, microdeductions appeared in the tests review even in inductive and matrix reasoning. Their presence in inductive tasks does not seem to be predictable or systematic, rather they could be taken as heuristic "deductions".

Regarding the cerebral review, the measures of neural correlates of deductive reasoning quantify changes in electrical activity (EEG, MEG, tDCS) and blood flux (fMRI) which come with deductive inferences or alternatively quantify the consequences of casual or reversible lesions (rTMS). Neural research contexts impose strict temporal frames in miliseconds and require multiple repetitions to obtain perceptible and significative statistical brain patterns. For these reasons neural and psychometrical measures are methodologically divergent. Moreover, neural research has defined deduction as "the ability to reach logical conclusions on the basis of prior information" (Goel et al, 2017) in sharp contrast with the CDC notion of deduction previously presented. Therefore, from a conceptual point of view, the neural deductive construct does not present the same rough inconsistencies as the psychometrical construct. However, there remains a more subtle consistency problem with the neural deductive construct, since experimental designs (Reverberi et al, 2010) exclusively consider integrable inferences and there seems to be no explicit knowledge of the difference between integrable and deductively valid inferences. For example, Goel (Goel, 2007; Goel et al, 2006), takes as the control task (hence non-deductive) non integrable sequences of statements, even if they are deductively valid. Practically all neural experiments employ exclusively integrated inferences which share variables. But classically valid deductive inferences are not constrained by integration and do not satisfy in general VSP (Robles and Méndez, 2011; Robles et al, 2012) What is measured then by experiments assessing VSP deductions with EEG, MEG, fMRI, PET and rTMS? Certainly not valid deduction (as most authors say), nor inferences "whose conclusions are certain if their premises are" (Evans et al, 2015; Méndez et al, 2012; Monti and Osherson, 2012; Monti et al, 2012a), nor "conclusions are necessary if their premises are" (Reverberi et al, 2007) since they are in fact constrained to integrated or VSP inferences.

This is not only a theoretical problem where validity is understood (confused) in these experiments with truth+meaning+relevance preserving consequence, since it is not casual that significant experimental differences correspond to divergent criteria of integration: compare for example (Goel et al, 2009) with (Parsons and Osherson, 2001). The presence of common terms or words among premises is sometimes taken as a criterion (syntactical criterion), in other occasions semantical or contextual criteria are adopted in distinct experimental studies like (Gerken, 2012; Santamaría et al, 2013).

This review has been limited to factual measures with which deductive inferences are assessed in psychological and neural contexts. Some subjective measures such as feeling of rightness or confidence have not been included. Specific semantical treatments of deductive validity (truth functional, probabilistic, constructive, game-theoretic, erotetic) are not explicitly distinguished. In the future, there are three alternative landscapes for factive deduction: either (1) some (sub)set of operationalized factive variables is present in all deductive processes, or (2) only some deductive features are factive, or (3) deductions are purely normative events and there are in fact no deductions.

References

Adams EW (1996) Four probability-preserving properties of inferences. Journal of Philosophical Logic 25(1):1–24

Alemany FS (1998) Hacia la lógica plástica: emergencia de la lógica del razonamiento visual. Contextos (31):281–296

Alvarez JA, Emory E (2006) Executive function and the frontal lobes: a meta-analytic review. Neuropsychology review 16(1):17–42

Alvarez Merino P, Requena Hernandez C, Salto Alemany F (2018) Measurement variables for deductive reasoning. REVISTA IBEROAMERICANA DE DIAGNOSTICO Y EVALUACION-E AVALIACAO PSICOLOGICA 4(49):59–75

Baggio G, Van Lambalgen M, Hagoort P (2012) The processing consequences of compositionality. In: The Oxford handbook of compositionality, Oxford University Press, pp 655–672

Baggio G, van Lambalgen M, Hagoort P (2015) Logic as Marr's computational level: Four case studies. Topics in cognitive science 7(2):287–298

Baggio G, Cherubini P, Pischedda D, Blumenthal A, Haynes JD, Reverberi C (2016) Multiple neural representations of elementary logical connectives. NeuroImage 135:300–310

Boolos GS, Burgess JP, Jeffrey RC (2002) Computability and logic. Cambridge university press

del Carmen Crivello M, Macbeth G, Fioramonti M, Razumiejczyk E (2016) Diferencias individuales en razonamiento deductivo: una revisión narrativa. Pensando Psicología 12(19):23–38

Carroll JB, et al (1993) Human cognitive abilities: A survey of factor-analytic studies. Cambridge University Press

Church A (1940) A formulation of the simple theory of types. The journal of symbolic logic 5(2):56–68

Cocchi L, Halford GS, Zalesky A, Harding IH, Ramm BJ, Cutmore T, Shum DH, Mattingley JB (2013) Complexity in relational processing predicts changes in functional brain network dynamics. Cerebral Cortex 24(9):2283–2296

Coetzee JP, Monti MM (2018) At the core of reasoning: Dissociating deductive and non-deductive load. Human brain mapping 39(4):1850–1861

Colberg M, Nester M, Trattner M (1985) Convergence of the inductive and deductive models in the measurement of reasoning abilities. Journal of Applied Psychology 70(4):681

Dehn MJ (2017) How working memory enables fluid reasoning. Applied Neuropsychology: Child 6(3):245–247

Dogramaci S (2017) Why is a valid inference a good inference? Philosophy and Phenomenological Research 94(1):61–96

Dove IJ (2016) Visual scheming: Assessing visual arguments. Argumentation and Advocacy 52(4):254–264

Dunn JM (2015) The relevance of relevance to relevance logic. In: Indian Conference on Logic and Its Applications, Springer, pp 11–29

Elosúa P, Almeida L (2016) BPR Batería de Pruebas de Razonamiento. TEA, Madrit

Evans JSB (2002) Logic and human reasoning: An assessment of the deduction paradigm. Psychological bulletin 128(6):978

Evans JSB, Over DE (2013) Reasoning to and from belief: Deduction and induction are still distinct. Thinking & Reasoning 19(3-4):267–283

Evans JSB, Thompson VA, Over DE (2015) Uncertain deduction and conditional reasoning. Frontiers in Psychology 6:398

Gerken M (2012) Univocal reasoning and inferential presuppositions. Erkenntnis 76(3):373–394

Gödel K (1930) Die vollständigkeit der axiome des logischen funktionenkalküls. Monatshefte für Mathematik und Physik 37(1):349–360

Gödel K (1995) Collected works: Volume III: Unpublished essays and lectures vol. 3

Goel V (2007) Anatomy of deductive reasoning. Trends in cognitive sciences 11(10):435–441

Goel V, Tierney M, Sheesley L, Bartolo A, Vartanian O, Grafman J (2006) Hemispheric specialization in human prefrontal cortex for resolving certain and uncertain inferences. Cerebral cortex 17(10):2245–2250

Goel V, Stollstorff M, Nakic M, Knutson K, Grafman J (2009) A role for right ventrolateral prefrontal cortex in reasoning about indeterminate relations. Neuropsychologia 47(13):2790–2797

Goel V, Navarrete G, Noveck IA, Prado J (2017) The reasoning brain: the interplay between cognitive neuroscience and theories of reasoning. Frontiers in human neuroscience 10:673

Hagoort P (2013) Muc (memory, unification, control) and beyond. Frontiers in psychology 4:416

Harman G (1984) Logic and reasoning. In: Foundations: Logic, Language, and Mathematics, Springer, pp 107–127

Harman G (2002) Internal critique: A logic is not a theory of reasoning and a theory of reasoning is not a logic. In: Studies in logic and practical reasoning, vol 1, Elsevier, pp 171–186

Heit E (2015) Brain imaging, forward inference, and theories of reasoning. Frontiers in human neuroscience 8:1056

Heit E, Rotello CM (2010) Relations between inductive reasoning and deductive reasoning. Journal of Experimental Psychology: Learning, Memory, and Cognition 36(3):805

Hintikka J (1999) Is logic the key to all good reasoning? In: Inquiry as inquiry: A logic of scientific discovery, Springer, pp 1–24

Hintikka J (2002) Hyperclassical logic (aka if logic) and its implications for logical theory. Bulletin of Symbolic Logic 8(3):404–423

Houdé O, Borst G (2014) Measuring inhibitory control in children and adults: brain imaging and mental chronometry. Frontiers in psychology 5:616

Houdé O, Borst G (2015) Evidence for an inhibitory-control theory of the reasoning brain. Frontiers in human neuroscience 9:148

Johnson G (2017) Argument and inference: An introduction to inductive logic. Mit Press

Johnson-Laird P, Khemlani SS (2013) Toward a unified theory of reasoning. In: Psychology of learning and motivation, vol 59, Elsevier, pp 1–42

Johnson-Laird PN (2008) Mental models and deductive reasoning. Reasoning: studies in human inference and its foundations pp 206–222

Kennedy J (2017) Turing, gödel and the "bright abyss". In: Philosophical Explorations of the Legacy of Alan Turing, Springer, pp 63–91

Khemlani S, Johnson-Laird PN (2012) Theories of the syllogism: A meta-analysis. Psychological bulletin 138(3):427

Koralus P, Mascarenhas S (2013) The erotetic theory of reasoning: bridges between formal semantics and the psychology of deductive inference. Philosophical Perspectives 27:312–365

Kyllonen PC, Christal RE (1990) Reasoning ability is (little more than) working-memory capacity?! Intelligence 14(4):389–433

Lassiter D, Goodman ND (2015) How many kinds of reasoning? inference, probability, and natural language semantics. Cognition 136:123–134

Liang P, Goel V, Jia X, Li K (2014) Different neural systems contribute to semantic bias and conflict detection in the inclusion fallacy task. Frontiers in human neuroscience 8:797

Magnani L (2011) Abduction, reason and science: Processes of discovery and explanation. Springer Science & Business Media

Markovits H, Brisson J, de Chantal PL, Thompson VA (2017) Interactions between inferential strategies and belief bias. Memory & cognition 45(7):1182–1192

Marr D (1982) A computational investigation into the human representation and processing of visual information. WH San Francisco: Freeman and Company

Martin-Löf P (1975) An intuitionistic theory of types: Predicative part. In: Studies in Logic and the Foundations of Mathematics, vol 80, Elsevier, pp 73–118

Mathieu R, Booth JR, Prado J (2015) Distributed neural representations of logical arguments in school-age children. Human brain mapping 36(3):996–1009

Méndez JM, Robles G, Salto F (2015) Brady's deep relevant logic dr plus the qualified factorization principles has the deep relevant condition. Logique et Analyse 232:547–565

Merino PÁ, Hernández CR, Alemany FS (2016) La integración más que la edad influye en el rendimiento del razonamiento deductivo. International Journal of Developmental and Educational Psychology Revista INFAD de Psicología 1(2):221–228

Monti MM, Osherson DN (2012) Logic, language and the brain. Brain research 1428:33–42

Monti MM, Osherson DN, Martinez MJ, Parsons LM (2007) Functional neuroanatomy of deductive inference: a language-independent distributed network. Neuroimage 37(3):1005–1016

Monti MM, Parsons LM, Osherson DN (2012a) Response to tzourio-mazoyer and zago: yes, there is a neural dissociation between language and reasoning. Trends in cognitive sciences 16(10):495–496

Monti MM, Parsons LM, Osherson DN (2012b) Thought beyond language: Neural dissociation of algebra and natural language. Psychological science 23(8):914–922

Morra S, Panesi S, Traverso L, Usai MC (2018) Which tasks measure what? reflections on executive function development and a commentary on podjarny, kamawar, and andrews (2017). Journal of experimental child psychology 167:246–258

Méndez J, Robles G, Salto F (2012) A general class of logical matrices for the variable-sharing property. Bulletin of Symbolic Logic 18:457

Oaksford M (2015) Imaging deductive reasoning and the new paradigm. Frontiers in human neuroscience 9:101

Oaksford M, Chater N (2007) Bayesian rationality: The probabilistic approach to human reasoning. Oxford University Press

O'Brien DP, Manfrinati A (2010) The mental logic theory of conditional proposition. Cognition and conditionals: Probability and Logic in Human Thinking pp 39–54

Parsons LM, Osherson D (2001) New evidence for distinct right and left brain systems for deductive versus probabilistic reasoning. Cerebral Cortex 11(10):954–965

Prado J, Van Der Henst JB, Noveck IA (2010) Recomposing a fragmented literature: How conditional and relational arguments engage different neural systems for deductive reasoning. Neuroimage 51(3):1213–1221

Prado J, Chadha A, Booth JR (2011) The brain network for deductive reasoning: a quantitative meta-analysis of 28 neuroimaging studies. Journal of cognitive neuroscience 23(11):3483–3497

Prado J, Mutreja R, Booth JR (2012) Fractionating the Neural Substrates of Transitive Reasoning: Task-Dependent Contributions of Spatial and Verbal Representations. Cerebral Cortex 23(3):499–507, DOI 10.1093/cercor/bhr389, URL `https://doi.org/10.1093/cercor/bhr389`, `http://oup.prod.sis.lan/cercor/article-pdf/23/3/499/882813/bhr389.pdf`

Prowse Turner JA, Thompson VA (2009) The role of training, alternative models, and logical necessity in determining confidence in syllogistic reasoning. Thinking & Reasoning 15(1):69–100

Ragni M, Knauff M (2013) A theory and a computational model of spatial reasoning with preferred mental models. Psychological review 120(3):561

Reverberi C, Cherubini P, Rapisarda A, Rigamonti E, Caltagirone C, Frackowiak RS, Macaluso E, Paulesu E (2007) Neural basis of generation of conclusions in elementary deduction. Neuroimage 38(4):752–762

Reverberi C, Cherubini P, Frackowiak RS, Caltagirone C, Paulesu E, Macaluso E (2010) Conditional and syllogistic deductive tasks dissociate functionally during premise integration. Human brain mapping 31(9):1430–1445

Reverberi C, Görgen K, Haynes JD (2012) Compositionality of rule representations in human prefrontal cortex. Cerebral cortex 22(6):1237–1246

Rips LJ (1994) The psychology of proof: Deductive reasoning in human thinking. Mit Press

Rips LJ (2001) Two kinds of reasoning. Psychological Science 12(2):129–134

Robles G, Méndez JM (2011) A class of simpler logical matrices for the variable-sharing property. Logic and Logical Philosophy 20(3):241–249

Robles G, Méndez JM, et al (2012) A general characterization of the variable-sharing property by means of logical matrices. Notre Dame Journal of Formal Logic 53(2):223–244

Rotello CM, Heit E (2009) Modeling the effects of argument length and validity on inductive and deductive reasoning. Journal of Experimental Psychology: Learning, Memory, and Cognition 35(5):1317

Routley R, Meyer RK (1972) The semantics of entailment: Iii. Journal of philosophical logic pp 192–208

Santamaría C, Tse PP, Moreno-Ríos S, García-Madruga JA (2013) Deductive reasoning and metalogical knowledge in preadolescence: a mental model appraisal. Journal of Cognitive Psychology 25(2):192–200

Schaeken W (2007) The mental models theory of reasoning: Refinements and extensions. Psychology Press, New York

Schechter J (2013) Deductive reasoning. The Encyclopedia of the Mind

Schroeders U, Wilhelm O (2010) Testing reasoning ability with handheld computers, notebooks, and paper and pencil. European Journal of Psychological Assessment

Schroyens W, Schaeken W, Handley S (2003) In search of counter-examples: Deductive rationality in human reasoning. The Quarterly Journal of Experimental Psychology Section A 56(7):1129–1145

Seisdedos N (2002) Rp-30: Resolución de problemas. Madrid: TEA Ediciones

Shimojima A (2015) Semantic properties of diagrams and their cognitive potentials. Center for the Study of Language and Information

Shye S (1988) Inductive and deductive reasoning: A structural reanalysis of ability tests. Journal of Applied Psychology 73(2):308

Shynkaruk JM, Thompson VA (2006) Confidence and accuracy in deductive reasoning. Memory & cognition 34(3):619–632

Singmann H, Klauer KC (2011) Deductive and inductive conditional inferences: Two modes of reasoning. Thinking & reasoning 17(3):247–281

Singmann H, Klauer KC, Beller S (2016) Probabilistic conditional reasoning: Disentangling form and content with the dual-source model. Cognitive Psychology 88:61–87

Skovgaard-Olsen N, Singmann H, Klauer KC (2016) The relevance effect and conditionals. Cognition 150:26–36

Stenning K, Van Lambalgen M (2012) Human reasoning and cognitive science. MIT Press

Stephens RG, Dunn JC, Hayes BK (2018) Are there two processes in reasoning? the dimensionality of inductive and deductive inferences. Psychological Review 125(2):218

Thompson VA, Turner JAP, Pennycook G (2011) Intuition, reason, and metacognition. Cognitive psychology 63(3):107–140

Tobin KG, Capie W (1981) The development and validation of a group test of logical thinking. Educational and Psychological measurement 41(2):413–423

Tsujii T, Watanabe S (2010) Neural correlates of belief-bias reasoning under time pressure: a near-infrared spectroscopy study. Neuroimage 50(3):1320–1326

Wechsler D (2008) Wechsler adult intelligence scale–fourth edition (wais–iv). San Antonio, TX: NCS Pearson 22:498

Wendelken C (2015) Meta-analysis: how does posterior parietal cortex contribute to reasoning? Frontiers in human neuroscience 8:1042

Wilhelm O (2005) Measuring reasoning ability. Handbook of understanding and measuring intelligence pp 373–392

Williamson T (2003) Blind reasoning. In: Aristotelian Society Supplementary Volume, Wiley Online Library, vol 77, pp 249–293

Wójcicki R (2013) Theory of logical calculi: basic theory of consequence operations, vol 199. Springer Science & Business Media

Żyluk N, Michta M, Urbański M (2018) Yet another shade of deduction. on measuring deductive flexibility and how it may relate to other cognitive abilities. Logic and Logical Philosophy 27(4):517–543

Cognitive Principles and Individual Differences in Human Syllogistic Reasoning

Emmanuelle-Anna Dietz Saldanha and Richard Mörbitz

Abstract Various psychological experiments have confirmed that humans do not reason according to Classical Logic. Instead of assuming that logic-based approaches in general might not be suitable for understanding and modeling human reasoning, we are convinced that logic can help us better understand why humans draw certain conclusions. However, we claim that Classical Logic is not adequate for this purpose. Recently, Khemlani and Johnson-Laird provided a meta-study on human syllogistic reasoning. The results of this meta-study based on six experiments are disappointing: Humans do not only systematically deviate from the assertions in Classical Logic, but from any of the 12 proposed cognitive theories. Surprisingly a computational logic approach, the Weak Completion Semantics, outperformed these theories. Yet, a major drawback, is that the evaluation was only carried out with respect to the aggregated data, i.e. all theories model only *the* human reasoner. However, an adequate cognitive theory should aim at modeling individual differences among reasoners, if they are observed. In this paper, we will re-assess a novel approach based on the Weak Completion Semantics, Clustering by Principles, that addresses these differences, and evaluate the results with respect to a non-aggregated dataset.

Key words: human syllogistic reasoning, logic programming, three-valued Łukasiewicz logic, individual reasoners

---

Emmanuelle-Anna Dietz Saldanha
International Center for Computational Logic
TU, Dresden, Germany
e-mail: emmanuelle.dietz@tu-dresden.de

Richard Mörbitz
International Center for Computational Logic
TU, Dresden, Germany
e-mail: richard.moerbitz@tu-dresden.de

(The authors are mentioned in alphabetical order)

1 Introduction

Even though the main logical formalism of Classical Logic (CL) has successfully been applied to our scientific reasoning, Kotarbiński (1913) already stated in 1913 *the need to revise the two-valued logic as it seemed to interfere with the freedom of human thinking* (Malinowski, 1993, p.17). In the last decades, cognitive scientists have repeatedly shown through experimental studies that human reasoning systematically deviates from conclusions that are valid under CL (cf. Wason (1968); Byrne (1989)), and thus they seem to agree with Kotarbiński: Classical (two-valued) logic is not an adequate candidate for a cognitive theory. But what should be the properties of such a candidate in the first place?

Cognitive theories that try to explain human reasoning need to be evaluated by their cognitive adequacy (Strube, 1992) and their performance should be compared to other theories. Conceptual cognitive adequacy describes to what extent a system corresponds to human conceptual knowledge, whereas inferential cognitive adequacy indicates whether the reasoning process of a system is structured similarly to the way humans reason (Knauff et al, 1997). In order to determine the required factors that are necessary for the evaluation of an adequate cognitive theory, we take the suggestion by Ragni and Stolzenburg (2015) as a starting point: We aim for a theory in which (i) all processes are comprehensible, (ii) the underlying assumptions represent general cognitive processes, and (iii) the predictions of our theory are evaluated with respect to a wide range of human reasoning tasks.

The results of the meta-study on syllogistic reasoning by Khemlani and Johnson-Laird (2012) show that the responses given by humans do not only systematically differ from the valid conclusions under CL, but from any other proposed cognitive theory (cf. Byrne and Johnson-Laird (1989); Johnson-Laird (1983); Rips (1994); Polk and Newell (1995); Chater and Oaksford (1999)).

Syllogisms originate from Aristotle (Barnes, 1984) and they are interesting because on the one hand they are easy to understand and on the other hand they are complex enough to require actual reasoning. A syllogism consists of two premises and one conclusion, each of which is a statement about properties over terms which can have one of the classical quantifiers in Table 1 (called *moods*). The pair of syllogistic premises can have four different orders as shown in Table 2 (called *figures*). Based on these moods and figures, 64 different pairs of syllogistic premises can be constructed.

A conclusion then consists of an assertion about two properties over terms where each of these properties occurs in exactly one of the premises, and there is a third property connecting both premises which does not occur in the conclusion.

For instance, consider the following pair of syllogistic premises:

Some bakers are artists.   *No chemists are bakers.*   (IE2)

Given these two premises, which conclusion on the relation between *artists* and *chemists* necessarily follows? According to Khemlani and Johnson-Laird (2012), the majority of participants in experimental studies concluded *No chemists are artists* (29 %) and *No valid conclusion* (27 %). The valid conclusion under CL, *Some artists*

Table 1: The classical moods and their formalization

Mood	First-order Logic	Short
affirmative universal	$\forall X(a(X) \rightarrow b(X))$	Aab
affirmative existential	$\exists X(a(X) \wedge b(X))$	Iab
negative universal	$\forall X(a(X) \rightarrow \neg b(X))$	Eab
negative existential	$\exists X(a(X) \wedge \neg b(X))$	Oab

Table 2: The classical figures

Figure	Premise 1	Premise 2
1	$a-b$	$b-c$
2	$b-a$	$c-b$
3	$a-b$	$c-b$
4	$b-a$	$b-c$

are not chemists, was concluded by only 16 %, which is exactly the threshold that the response had not been chosen randomly (Khemlani and Johnson-Laird, 2012). Usually in these experiments, participants are supposed to give only one response, thus all three responses above were given by different participants.

A major drawback of most cognitive theories is that only the participants' aggregated data was considered: These theories did not account for differences among individuals, i.e. they only considered the data summed up with respect to the individual response possibilities. In contrast, non-aggregated data records the individual participants' response patterns, i.e. the complete set of responses that are given by each participant.

As the above example and recent investigations on psychological experiments show (cf. Johnson-Laird and Khemlani, 2016; Ragni et al, 2017), it seems that different humans represent and reason about conditionals differently. This implies that *the human reasoner does not exist*, instead there seem to exist various groups of reasoners, i.e. reasoning clusters. The identification of these differences and the specification of these clusters is central for the development of an adequate cognitive theory. Only recently approaches have been proposed that account for individual differences in syllogistic reasoning (cf. Johnson-Laird and Khemlani, 2016; Dietz Saldanha et al, 2017).

Here, we provide a qualitative assessment of the *clustering by principles* approach proposed in (Dietz Saldanha et al, 2017), that performed well on aggregated data, and investigate whether these clusters are also adequate with respect to the non-aggregated dataset provided by Johnson-Laird and Khemlani (2016). In Section 2 and 3, we introduce the underlying computational logic approach, the cognitive principles and their formalization. Section 4.1 illustrates an example and briefly discusses the performance of the approach presented in (Dietz Saldanha et al, 2017) on non-aggregated data. Section 4.2 provides a novel characterization of the individuals' reasoning patterns through cognitive principles.

2 Clustering by Principles

Clustering by principles was first presented in (Dietz Saldanha et al, 2017), which is an extension of the computational logic approach for human syllogistic reasoning presented in (Costa et al, 2017b). The authors in (Dietz Saldanha et al, 2017) conjectured that reasoning clusters can be expressed through cognitive principles and heuristic strategies. Cognitive principles are assumptions (not necessarily valid under CL) made by humans while reasoning, whereas heuristic strategies describe the methods that are applied when conclusions are not derived by reasoning but instead based on structural patterns or biases: Three reasoning clusters and two heuristic strategies were specified, and achieved a match of 92 %, which was the highest result compared to the presented approaches by Khemlani and Johnson-Laird (2012). Figure 1 illustrates the components of the system: First, groups of reasoners were specified through sets of underlying cognitive principles and heuristic strategies. After that, these sets were translated into logic programs (Section 3.1-3.2), representing the reasoning clusters. Next, the respective models were computed (Section 2.1-2.3). After that, for each cluster, the entailed relations between a and c were extracted (Section 3.3). The final response distribution was specified according to the number of reasoning clusters that entailed the conclusion. Finally, an illustrative representation of these clusters with respect to their distribution was done by means of multinomial process trees (Riefer and Batchelder, 1988).

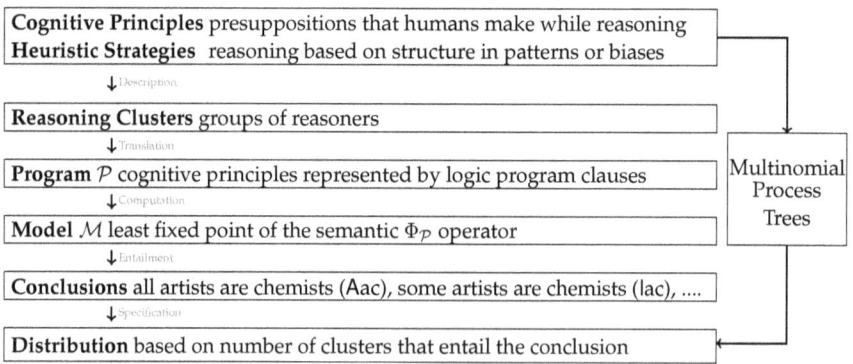

Fig. 1: The methodology of *clustering by principles*

The suggested computational logic approach for the formal representation and reasoning of these clusters, is the Weak Completion Semantics (WCS) first introduced by Hölldobler and Kecana Ramli (2009) which has shown to model a wide range of human reasoning tasks (Hölldobler, 2015) and will be introduced in the following.

2.1 Contextual Logic Programs

We introduce an extension of WCS first presented in (Dietz Saldanha et al, 2017). *Contextual (logic) program clauses* are expressions of the forms $A \leftarrow L_1 \wedge \ldots \wedge L_m \wedge \mathsf{ctxt}(L_{m+1}) \wedge \ldots \wedge \mathsf{ctxt}(L_{m+p})$ (called *rules*), $A \leftarrow \top$ (called *facts*), $A \leftarrow \bot$ (called *negative assumptions*)[1] and $A \leftarrow \mathsf{U}$ (called *unknown assumptions*), where A is an atom and L_i with $1 \leq i \leq m+p$, is a literal. A is called *head* and $L_1 \wedge \ldots \wedge L_m \wedge \mathsf{ctxt}(L_{m+1}) \wedge \ldots \wedge \mathsf{ctxt}(L_{m+p})$ as well as $\top, \bot$ and U, standing for *true*, *false* and *unknown* respectively, are called *body* of the corresponding clauses. A *contextual program*, denoted by $\mathcal{P}$, is a finite set of contextual program clauses. We restrict terms to be either constants or variables only, hence we consider so-called *data logic programs*. If a clause contains variables, then they are implicitly universally quantified within the scope of the entire clause. A *ground clause* is a clause that does not contain variables. $\mathsf{g}\mathcal{P}$ denotes the set of all ground clauses of $\mathcal{P}$. An atom A is *defined in* $\mathsf{g}\mathcal{P}$ iff $\mathsf{g}\mathcal{P}$ contains a clause with head A. The *definition of A in $\mathsf{g}\mathcal{P}$* is defined as $\mathsf{def}(A, \mathcal{P}) = \{A \leftarrow Body \mid A \leftarrow Body \text{ is a clause occurring in } \mathsf{g}\mathcal{P}\}$. A is *undefined in* $\mathsf{g}\mathcal{P}$ iff A is not defined in $\mathsf{g}\mathcal{P}$. The set of all atoms that are undefined in $\mathsf{g}\mathcal{P}$ is denoted by $\mathsf{undef}(\mathcal{P})$. Consider the following transformation for a given $\mathcal{P}$: 1. For each atom A defined in $\mathsf{g}\mathcal{P}$, replace all clauses of the form $A \leftarrow Body_1, \ldots, A \leftarrow Body_m$ occurring in $\mathsf{g}\mathcal{P}$ by $A \leftarrow Body_1 \vee \ldots \vee Body_m$. 2. Replace all occurrences of $\leftarrow$ by $\leftrightarrow$. The obtained ground set of equivalences is called the *weak completion* of $\mathcal{P}$. In the sequel, we omit the word *contextual* when we refer to programs.

2.2 Three-Valued Łukasiewicz Logic Extended with ctxt

We consider the three-valued Łukasiewicz (1920) logic extended with the ctxt connective, for which the corresponding truth values are $\top, \bot$ and U, meaning *true*, *false* and *unknown*, respectively. A *three-valued interpretation* I is a mapping from the set of logical formulas to the set of truth values $\{\top, \bot, \mathsf{U}\}$, represented as a pair $I = \langle I^\top, I^\bot \rangle$ of two disjoint sets of atoms: $I^\top = \{A \mid A \text{ is mapped to } \top \text{ under } I\}$ and $I^\bot = \{A \mid A \text{ is mapped to } \bot \text{ under } I\}$. Atoms which do not occur in $I^\top \cup I^\bot$ are mapped to U. The truth value of a given formula under I is determined according to the truth tables in Table 3.

---

[1] Under WCS, the negative assumption will become $A \leftrightarrow \bot$, which, however, can be overwritten by other rules and facts (*defeating* the assumption).

Table 3: The truth tables for the connectives under the three-valued Łukasiewicz logic and for ctxt. L is a literal and $\top$, $\bot$, and U denote *true*, *false*, and *unknown*, respectively

F	$\neg F$		$\wedge$	$\top$	U	$\bot$		$\vee$	$\top$	U	$\bot$		$\leftarrow$	$\top$	U	$\bot$		$\leftrightarrow$	$\top$	U	$\bot$		L	ctxt(L)[2]
$\top$	$\bot$		$\top$	$\top$	U	$\bot$		$\top$	$\top$	$\top$	$\top$		$\top$	$\top$	$\top$	$\top$		$\top$	$\top$	U	$\bot$		$\top$	$\top$
$\bot$	$\top$		U	U	U	$\bot$		U	$\top$	U	U		U	U	$\top$	$\top$		U	U	$\top$	U		$\bot$	$\bot$
U	U		$\bot$	$\bot$	$\bot$	$\bot$		$\bot$	$\top$	U	$\bot$		$\bot$	$\bot$	U	$\top$		$\bot$	$\bot$	U	$\top$		U	$\bot$

$I(F) = \top$ denotes that a formula F is mapped to true under I. A *three-valued model* $\mathcal{M}$ of $\mathcal{P}$ is a three-valued interpretation such that $\mathcal{M}(A \leftarrow Body) = \top$ for each $A \leftarrow Body \in \mathsf{g}\mathcal{P}$. Let $I = \langle I^\top, I^\bot \rangle$ and $J = \langle J^\top, J^\bot \rangle$ be two interpretations. $I \subseteq J$ iff $I^\top \subseteq J^\top$ and $I^\bot \subseteq J^\bot$. I is the *least model* of $\mathcal{P}$ iff for any other model J of $\mathcal{P}$ it holds that $I \subseteq J$.

A set of *integrity constraints* $\mathcal{IC}$ consists of clauses of the form $\mathsf{U} \leftarrow Body$, where $Body$ is a conjunction of literals. An interpretation maps an integrity constraint to $\top$ iff $Body$ is either mapped to U or $\bot$. Given an interpretation I and a set of integrity constraints $\mathcal{IC}$, I satisfies $\mathcal{IC}$ iff all clauses in $\mathcal{IC}$ are true under I.

2.3 Reasoning with Respect to the Least Fixed Point of $\Phi_{\mathcal{P}}$

Consider the following semantic operator introduced by Stenning and van Lambalgen (2008): Let $I = \langle I^\top, I^\bot \rangle$ be an interpretation. $\Phi_{\mathcal{P}}(I) = \langle J^\top, J^\bot \rangle$, where

$$J^\top = \{A \mid A \leftarrow Body \in \mathrm{def}(A, \mathcal{P}) \text{ and } Body \text{ is } true \text{ under } \langle I^\top, I^\bot \rangle\}$$
$$J^\bot = \{A \mid \mathrm{def}(A, \mathcal{P}) \neq \varnothing \text{ and }$$
$$\qquad Body \text{ is } false \text{ under } \langle I^\top, I^\bot \rangle \text{ for all } A \leftarrow Body \in \mathrm{def}(A, \mathcal{P})\}.$$

The weak completion of non-contextual programs[3] always has a least model under Łukasiewicz logic, which corresponds to the least fixed point of $\Phi_{\mathcal{P}}$ (Hölldobler and Kencana Ramli, 2009). The interested reader is referred to (Dietz et al, 2014) for the correspondence to other logic programming approaches.

Different than for non-contextual programs, the weak completion of contextual programs might have more than one minimal model. Furthermore, if $\mathcal{P}$ is a contextual program, the least fixed point of $\Phi_{\mathcal{P}}$ might not even exist. However, for contextual programs without cycles, there always exists a unique supported model, which corresponds to the least fixed point of $\Phi_{\mathcal{P}}$ (Dietz Saldanha, Hölldobler, and Pereira, 2017; Dietz Saldanha, Hölldobler, and Philipp, 2017). This model is *supported* in the sense

---

[2] Thanks to the anonymous reviewer for the observation, that the ctxt operator seems to be a special case of Łukasiewicz necessity operator (cf. Malinowski, 1993, p. 20), restricted to literals.

[3] Non-contextual programs are programs that do not contain rules in which ctxt appears in the body of any rule, i.e. $p = 0$.

that it will always be computed by $\Phi_\mathcal{P}$, independent from the interpretation we start to iterate $\Phi_\mathcal{P}$ with. Let us denote this model as $\mathcal{M}_\mathcal{P}$. As the fixed point of $\Phi_\mathcal{P}$ is unique and only computes the supported minimal model of wc $\mathcal{P}$, we define $\mathcal{P} \models_{wcs} F$ iff $\mathcal{M}_\mathcal{P}(F) = \top$. For all programs considered in the sequel, $\mathcal{M}_\mathcal{P}$ exists.

3 Cognitive Principles

Several principles for quantified statements have been developed in (Costa et al, 2017b,a; Dietz Saldanha et al, 2017). In the following, the programs are specified using the predicates *y* and *z*, where *yz* can be replaced by *ab*, *ba*, *cb*, or *bc* and depend on the figures in Table 2. First, we introduce the basic principles that are assumed for all premises when applicable. In Section 3.2, we present the advanced principles, which will later be used to distinguish between the clusters of reasoners. A complete specification of the logic programs with respect to the moods according to all principles can be found in (Dietz Saldanha et al, 2017).

3.1 Basic Principles

3.1.1 Conditional Representation (conditional) and License for Inference (license)

Consider *All y are z* (Ayz), which in CL, is represented as $\forall X(y(X) \to z(X))$. Similarly, we assume that humans also understand this statement as a conditional sentence (conditional), but also defeasible. According to Stenning and van Lambalgen (2008), they assume a *license for the inference* and might allow exceptions to the conditional (license). For Ayz this can be explicitly expressed by *all y that are not abnormal are z*. By default, we assume that *nothing is abnormal*.

3.1.2 Existential Import (import)

Furthermore, according to Grice (1975), humans have a pragmatic understanding of quantifiers in the language. For instance, in natural language, we normally do not quantify over things that do not exist. Consequently, *All z* implies *Some z exist*. This is known as the *Gricean implicature* or the *existential import* (import), which also corresponds to the Aristotelian interpretation (Parry and Hacker, 1991) and is assumed by several cognitive theories (cf. Johnson-Laird, 1983; Rips, 1994). Together with the introduced cognitive principles in Section 3.1.1, the logic program representing Ayz, $\mathcal{P}_{Ayz}$, is as follows:

$$z(X) \leftarrow y(X) \wedge \neg ab_{yz}(X). \quad \text{(conditional \& license)}$$
$$ab_{yz}(X) \leftarrow \text{ctxt}(z'(X)). \quad \text{(license)}$$
$$y(o) \leftarrow \top. \quad \text{(import)}$$

Note that $z'(X)$ appears within the ctxt operator in the body of the second clause, where z' represents the negation of z (which will be introduced by the transformation principle in Section 3.1.3). Nothing is abnormal, unless the negation of z is true. We obtain $\mathcal{M}_{\mathcal{P}_{Ayz}} = \langle \{y(o), z(o)\}, \{ab_{yz}(o)\} \rangle$.

3.1.3 Negation by Transformation (transformation) and No Inference through Double Negation (noDoubleNeg)

Consider the statement *No y is z.* (**Eyz**): Similarly to the previous case, license and import apply here as well. The logic programs specified in Subsection 2.1 do not allow heads of clauses to be negative literals. Therefore, a negative conclusion $\neg p(X)$ is represented by introducing an auxiliary formula $p'(X)$ together with the clause $p(X) \leftarrow \neg p'(X)$ and the integrity constraint $\mathsf{U} \leftarrow p(X) \wedge p'(X)$ (transformation). This is a widely used technique in Logic Programming (cf. choice rules in Gebser et al, 2012). We also introduce the *no inference through double negation* principle (noDoubleNeg): Under the Weak Completion Semantics, a positive conclusion can be derived from two negations appearing in two conditionals. However, it seems that humans normally do not reason in such a way (Khemlani and Johnson-Laird, 2012). Hence, we block them with the help of abnormalities. $\mathcal{P}_{Eyz}$, the logic program representing **Eyz** is as follows:

$$z'(X) \leftarrow y(X) \wedge \neg ab_{ynz}(X). \quad \text{(conditional \& transformation \& license)}$$
$$ab_{ynz}(X) \leftarrow \text{ctxt}(z(X)). \quad \text{(license)}$$
$$z(X) \leftarrow \neg z'(X) \wedge \neg ab_{nzz}(X). \quad \text{(transformation \& license)}$$
$$y(o_1) \leftarrow \top. \quad \text{(import)}$$
$$ab_{nzz}(o_1) \leftarrow \bot. \quad \text{(license \& noDoubleNeg)}$$

As we now have both $z(X)$ and $z'(X)$ in the head of some clause, we need to add the following integrity constraints: $\mathsf{U} \leftarrow z(X) \wedge z'(X)$. We obtain

$$\mathcal{M}_{\mathcal{P}_{Eyz}} = \langle \{y(o_1), z'(o_1)\}, \{ab_{ynz}(o_1), ab_{nzz}(o_1), z(o_1)\} \rangle.$$

3.1.4 Converse Implication (converse) and Unknown Generalization (unknownGen)

Consider the statement *Some y are z.* (**Iyz**), which in CL, is represented as $\exists X(y(X) \wedge z(X))$ and which is semantically equivalent to $\exists X(z(X) \wedge y(X))$ (converse). Another observation is that humans seem to distinguish between *some y are z* and *all y are z* (cf. Grice, 1975; Khemlani and Johnson-Laird, 2012). Accordingly, if we only observe

Cognitive Principles and Individual Differences in Human Syllogistic Reasoning

that an object o belongs to y and z then we do not want to conclude both, *some y are z* and *all y are z*. In order to distinguish between *some y are z* and *all y are z*, we introduce the *unknown generalization* principle (unknownGen): If we know that *some y are z*, then there must not only be an object o_1 which belongs to y and z, but there must be another object o_2 which belongs to y and which does not belong to z. $\mathcal{P}_{Iyz}$, the logic program representing Iyz, is as follows:[4]

$$z(X) \leftarrow y(X) \land \neg ab_{yz}(X). \qquad \text{(conditional \& license)}$$
$$ab_{yz}(o_1) \leftarrow \bot. \qquad \text{(unknownGen \& license)}$$
$$y(o_1) \leftarrow \top. \qquad \text{(import)}$$
$$y(o_2) \leftarrow \top. \qquad \text{(unknownGen)}$$

Consider the second clause: Different to the universal moods, **E** and **A**, ab_{yz} is only false for o_1, which in turn implies that ab_{yz} is unknown for all other objects in the program. We obtain $\mathcal{M}_{\mathcal{P}_{Iyz}} = \langle \{y(o_1), y(o_2), z(o_1)\}, \{ab_{yz}(o_1)\} \rangle$. One should observe that $ab_{yz}(o_2)$ is an unknown assumption in $\mathcal{P}_{Iyz}$ and, hence, $\mathcal{M}_{\mathcal{P}_{Iyz}}(z(o_2)) = \mathsf{U}$.

Consider the statement *Some y are not z* (**O**yz), which in CL is represented as $\exists X(y(X) \land \neg z(X))$. As in the previous cases, conditional, license, and unknownGen are assumed. As the conclusion is again a negation of a claim, transformation and noDoubleNeg apply as well. $\mathcal{P}_{Oyz}$, the logic program representing **O**yz is as follows:

$$z'(X) \leftarrow y(X) \land \neg ab_{ynz}(X). \qquad \text{(conditional \& transformation \& license)}$$
$$ab_{ynz}(o_1) \leftarrow \bot. \qquad \text{(license \& unknownGen)}$$
$$z(X) \leftarrow \neg z'(X) \land \neg ab_{nzz}(X). \qquad \text{(license \& transformation)}$$
$$y(o_1) \leftarrow \top. \qquad \text{(import)}$$
$$y(o_2) \leftarrow \top. \qquad \text{(unknownGen)}$$
$$ab_{nzz}(o_1) \leftarrow \bot. \qquad \text{(license \& noDoubleNeg)}$$
$$ab_{nzz}(o_2) \leftarrow \bot. \qquad \text{(license \& noDoubleNeg)}$$

We have to add the integrity constraint $\mathsf{U} \leftarrow z(X) \land z'(X)$ and obtain

$$\mathcal{M}_{\mathcal{P}_{Oyz}} = \langle \{y(o_1), y(o_2), z'(o_1)\}, \{ab_{ynz}(o_1), ab_{nzz}(o_1), ab_{nzz}(o_2), z(o_1)\} \rangle.$$

3.2 Advanced Principles

Different than the basic principles, we assume that the principles introduced now, do not apply for all reasoners.

---

[4] We omit the clauses introduced by (**converse**), as they are symmetric. They can be found in (Dietz Saldanha et al, 2017).

3.2.1 Contraposition (contra)

In CL, the conditional statement $\forall(X)(a(X) \rightarrow b(X))$, is logically equivalent to its contrapositive, $\forall(X)(\neg b(X) \rightarrow \neg a(X))$. There is evidence that some of the participants might directly or indirectly apply the contrapositive (contra) within a reasoning task (cf. Rips, 1994; O'Brien et al, 1994). Consider again $\mathcal{P}_{Ayz}$, the logic program representing Ayz from Section 3.1. The clauses that represent the contraposition principles are as follows:

$$y'(X) \leftarrow \neg z(X) \wedge \neg ab_{zny}(X). \qquad \text{(conditional \& license \& contra)}$$
$$ab_{zy}(X) \leftarrow \text{ctxt}(y(X)). \qquad \text{(license \& contra)}$$
$$y(X) \leftarrow \neg y'(X) \wedge \neg ab_{nyy}(X). \qquad \text{(transformation \& license \& contra)}$$

The application of contra for the E mood, *No y is z*, yields *No z is y*, and consists of the following clauses:

$$y'(X) \leftarrow z(X) \wedge \neg ab_{zny}(X). \qquad \text{(conditional \& license \& contra)}$$
$$ab_{zny}(X) \leftarrow \text{ctxt}(y(X)). \qquad \text{(license \& contra)}$$
$$y(X) \leftarrow \neg y'(X) \wedge \neg ab_{nyy}(X). \qquad \text{(transformation \& license \& contra)}$$

3.2.2 Generalization (generalization)

If all principles introduced so far are applied to a pair of existentially quantified premises (I or O), then the only object about which an inference can be made is the one resulting from import with respect to the premise it introduced (i.e. only inferences about the properties of the object about a and b or b and c). There is, however, evidence that some humans still draw conclusions about a and c in such circumstances, i.e. they seem to generalize (universally quantify) the existential premise with respect to other objects (generalization). Consider again $\mathcal{P}_{Iyz}$ of Section 3.1, where ab_{yz} is only false for o_1. In case generalization applies, then $\mathcal{P}_{Iyz}$ is extended by the following two clauses:

$$ab_{yz}(X) \leftarrow \text{ctxt}(z'(X)). \qquad \text{(license \& generalization)}$$
$$ab_{yz}(o_2) \leftarrow \mathsf{U}. \qquad \text{(license \& generalization)}$$

Similar to the case of $\mathcal{P}_{Ayz}$ for the representation of the basic principles of Ayz, the ctxt operator in the body of the first clause is false for all o for which $z'(o)$ is unknown. This clause transforms the representation for the existential quantification over Iyz in $\mathcal{P}_{Iyz}$ into a universal one. In order to maintain unknownGen, the second clause is added: $ab_{yz}(o_2) \leftarrow \mathsf{U}$ guarantees that $ab_{yz}(o_2)$ stays unknown, even though $\text{ctxt}(z'(o_2))$ might be false. Recall that programs are considered under their weak completion, i.e. if $\text{ctxt}(z'(o_2))$ is false, then

$$ab_{yz}(o_2) \leftrightarrow (\text{ctxt}(z'(o_2)) \vee \mathsf{U}) \equiv ab_{yz}(o_2) \leftrightarrow \mathsf{U}.$$

3.2.3 Search for Alternative Conclusions (searchAlt)

When participants are faced with NVC (*No valid conclusion*), they might not want to accept this conclusion and proceed to check whether there exists unknown information that might be relevant. This information may be explanations about facts introduced from either import or unknownGen. We use the first as source for observations, as they are used directly to infer new information. Consider the following pair of syllogistic premises:

All artists are bakers. Some bakers are not chemists. (AO1)

Under CL, *No valid conclusion* follows. However, 62 % of the participants concluded *Some artists are not chemists*. This conclusion can be modeled by abductive reasoning, where the system searches for explanations for a given observation (searchAlt). The second premise states that some bakers exist, i.e. the observation is $\mathcal{O} = \{b(o_2)\}$. $\mathcal{O}$ can be explained by the first premise, namely that this baker is an artist. Under WCS, this reasoning procedure was formalized by means of skeptical abduction. The interested reader is referred to (Costa et al, 2017b) for more details.

3.3 Entailment of Conclusions

We specify when $\mathcal{M}_\mathcal{P}$ entails a conclusion with respect to yz:[5]

Ayz (all) $\mathcal{P} \models Ayz$ iff there exists an object o such that $\mathcal{P} \models_{wcs} y(o)$ and for all objects o we find that if $\mathcal{P} \models_{wcs} y(o)$ then $\mathcal{P} \models_{wcs} z(o)$.
Eyz (no) $\mathcal{P} \models Eyz$ iff there exists an object o_1 such that $\mathcal{P} \models_{wcs} y(o_1)$ and for all objects o_1 we find that if $\mathcal{P} \models_{wcs} y(o_1)$ then $\mathcal{P} \models_{wcs} \neg z(o_1)$.[6]
Iyz (some) $\mathcal{P} \models Iyz$ iff there exists an object o_1 such that $\mathcal{P} \models_{wcs} y(o_1) \wedge z(o_1)$ and there exists an object o_2 such that $\mathcal{P} \models_{wcs} y(o_2)$ and $\mathcal{P} \not\models_{wcs} z(o_2)$ and there exists an object o_3 such that $\mathcal{P} \models_{wcs} z(o_3)$ and $\mathcal{P} \not\models_{wcs} y(o_3)$.
Oyz (some are not) $\mathcal{P} \models Oyz$ iff there exists an object o_1 such that $\mathcal{P} \models_{wcs} y(o_1) \wedge \neg z(o_1)$ and there exists an object o_2 such that $\mathcal{P} \models_{wcs} y(o_2)$ and $\mathcal{P} \not\models_{wcs} \neg z(o_2)$.
NVC (no valid conclusion) iff none of the previous conclusions can be derived.

For Iyz (some) the conversion is explicitly specified, motivated by the converse principle. By requiring for Iyz (Oyz, resp.) the existence of an object with the first property y, for which the second property z has to be either either false or unknown (true or unknown, resp.), Ayz and Iyz, (Eyz and Oyz, resp.), exclude each other.

---

[5] yz can be replaced by ac or ca.
[6] When the contraposition principles applies, then additionally the following condition needs to hold as well: There exists an object o_2 such that $\mathcal{P} \models_{wcs} z(o_2)$ and for all objects o_2 we find that if $\mathcal{P} \models_{wcs} z(o_2)$ then $\mathcal{P} \models_{wcs} \neg y(o_2)$.

4 Evaluation

4.1 Clusters

As illustration, let us consider IE2 from the introduction:

Some bakers are artists. No chemists are bakers. (IE2)

The majority of participants in experimental studies concluded *No chemists are artists* (29 %), *No valid conclusion* (27 %) and *Some artists are not chemists* (16 %). In order to account for the individual differences among participants, three clusters on principles and two on heuristic strategies have been suggested in (Dietz Saldanha et al, 2017). The two heuristic strategies are the atmosphere bias (Woodworth and Sells, 1935) and the matching bias (Wetherick and Gilhooly, 1995).[7] The three reasoning clusters have been characterized by the following principles:[8] (1) basic, converse for I, and searchAlt, (2) basic, converse for I, and generalization, and (3) basic, converse for I and E, and contra. This approach achieved a match of 92 % with the aggregated data reported in (Khemlani and Johnson-Laird, 2012), and was the highest result compared to the other approaches.

Table 4 summarizes the three clusters together with their conclusions for IE2.

Table 4: Clusters characterized by cognitive principles and entailments for IE2

Principle	Cluster (1)	Cluster (2)	Cluster (3)
conditional	✓	✓	✓
license	✓	✓	✓
import	✓	✓	✓
unknownGen	✓	✓	✓
converse for I	✓	✓	✓
searchAlt	✓	-	-
generalization	-	✓	-
converse for E	-	-	✓
contra	-	-	✓
	⇓	⇓	⇓
	NVC	*No chemists are artists*	*Some artists are not chemists*

The corresponding logic program $\mathcal{P}_{IE2}$ for cluster (1) consists of the following clauses:

---

[7] The interested reader is referred to (Dietz Saldanha et al, 2017) for details on their application in syllogistic reasoning within the Weak Completion Semantics.

[8] basic refers to the set of basic principles introduced in Subsection 3.1.

$$a(X) \leftarrow b(X) \land \neg ab_{ba}(X). \qquad \text{(conditional \& license)}$$
$$ab_{ba}(o_1) \leftarrow \bot. \qquad \text{(license \& unknownGen)}$$
$$b(o_1) \leftarrow \top. \qquad \text{(import)}$$
$$b(o_2) \leftarrow \top. \qquad \text{(unknownGen)}$$
$$b(X) \leftarrow a(X) \land \neg ab_{ab}(X). \qquad \text{(conditional \& license \& converse)}$$
$$ab_{ab}(o_3) \leftarrow \bot. \qquad \text{(unknownGen \& license \& converse)}$$
$$a(o_3) \leftarrow \top. \qquad \text{(import \& converse)}$$
$$a(o_4) \leftarrow \top. \qquad \text{(unknownGen \& converse)}$$
$$b'(X) \leftarrow c(X) \land \neg ab_{cnb}(X). \qquad \text{(license \& transformation)}$$
$$ab_{cnb}(X) \leftarrow \text{ctxt}(b(X)). \qquad \text{(license)}$$
$$b(X) \leftarrow \neg b'(X) \land \neg ab_{nbb}(X). \qquad \text{(license \& transformation)}$$
$$c(o_5) \leftarrow \top. \qquad \text{(import)}$$
$$ab_{nbb}(o_5) \leftarrow \bot. \qquad \text{(license \& noDoubleNeg)}$$

We obtain $\mathcal{M}_{\mathcal{P}_{\text{IE2}}} = \langle I^\top, I^\bot \rangle$, where

$I^\top = \{b(o_1), a(o_1), b(o_2), a(o_3), b(o_3), a(o_4), c(o_5), b'(o_5), ab_{cnb}(o_1), ab_{cnb}(o_2),$
$\quad ab_{cnb}(o_3)\},$
$I^\bot = \{ab_{ba}(o_1), ab_{ab}(o_3), ab_{cnb}(o_5), ab_{nbb}(o_5), ab_{cnb}(o_4)\},$

for which according to the entailment of conclusions as defined in Section 3.3, *No valid conclusion* follows.[9]

Applying the contraposition principle to *No chemists are bakers* yields *No bakers are chemists*. $\mathcal{P}_{\text{IE2}}^{\text{contra}}$ consists of the following clauses:

$$\mathcal{P}_{\text{IE2}} \cup \{c'(X) \leftarrow b(X) \land \neg ab_{bnc}(X), \quad \text{(basic \& license \& transformation \& contra)}$$
$$ab_{bnc}(X) \leftarrow \text{ctxt}(c(X)), \qquad \text{(contra \& license)}$$
$$c(X) \leftarrow \neg c'(X) \land \neg ab_{ncc}(X), \qquad \text{(license \& transformation \& contra)}$$
$$b(o_6) \leftarrow \top, \qquad \text{(import \& contra)}$$
$$ab_{ncc}(o_6) \leftarrow \bot\} \qquad \text{(license \& noDoubleNeg \& contra)}$$

We obtain $\mathcal{M}_{\mathcal{P}_{\text{IE2}}^{\text{contra}}} = \langle I^{\top'}, I^{\bot'} \rangle$ where

$I^{\top'} = I^\top \cup \{b(o_6), c'(o_3), c'(o_1), c'(o_2), c'(o_6), ab_{bnc}(o_5), ab_{cnb}(o_6)\},$
$I^{\bot'} = I^\bot \cup \{ab_{bnc}(o_i) \mid i \in \{1,2,3,4,6\}\}$
$\quad \cup \{c(o_i), b'(o_1) \mid i \in \{1,2,3,6\}\} \cup ab_{cnb}(o_3), ab_{ncc}(o_6)\}.$

Some artists are not chemists is entailed by $\mathcal{M}_{\mathcal{P}_{\text{IE2}}^{\text{contra}}}$ because $a(o_1)$ is true and $c(o_1)$ is false and $a(o_4)$ is true, but $c(o_4)$ is unknown.

---

[9] For simplicity we ignore the application of searchAlt, which is described in more detail in (Costa et al, 2017b).

Applying the generalization principle to *Some bakers are artists* and its converse, yields to *All bakers are artists* and *All artists are bakers*, respectively. $\mathcal{P}_{IE2}^{gen}$ consists of the clauses in $\mathcal{P}_{IE2}$ and the following ones:

$\mathcal{P}_{IE2} \cup \{ab_{ba}(X) \leftarrow \text{ctxt}(a'(X)), ab_{ba}(o_2) \leftarrow \text{U},$ (basic & license & generalization)
$ab_{ab}(X) \leftarrow \text{ctxt}(b'(X)), ab_{ab}(o_4) \leftarrow \text{U}\}.$ (license & generalization & converse)

We obtain $\mathcal{M}_{\mathcal{P}_{IE2}^{gen}} = \langle I^{\top''}, I^{\bot''} \rangle$ where

$I^{\top''} = I^{\top} \cup \{ab_{ab}(o_5)\},$
$I^{\bot''} = I^{\bot} \cup \{a(o_5), b(o5)\} \cup \{ab_{ab}(o_i) \mid i \in \{1, 2\}\}$
$\cup \{ab_{ba}(o_i) \mid i \in \{3, 4, 5\}\} \cup \{ab_{cnb}(o_3) \mid i \in \{1, 2, 3\}\}.$

No chemists are artists is entailed by $\mathcal{M}_{\mathcal{P}_{IE2}^{gen}}$, because for all objects which have the property c, (which only applies for o_5), they do not have the property a.

On aggregated data, the clusters performed fairly well with a match of 92 %. We are interested whether these clusters also correspond to the reasoning patterns observed among individuals. For this purpose, we investigated the non-aggregated dataset provided by Johnson-Laird and Khemlani (2016) consisting of the responses of 20 participants for two weeks, i.e. the participants carried out the task twice, with a break of one week. The results are two-fold:

Aggregated evaluation Most of the majority's responses can be explained by the clusters' predictions. In the first week the majority's conclusion for 63 out of 64 pair of syllogistic premises and in the second week all majority's conclusions were predicted by the clustering by principles approach, i.e. by cluster (1), (2) or (3) as specified in Table 4.

Non-aggregated evaluation None of the clusters fitted any individual participants' response pattern, i.e. none of the participants belonged to a certain cluster.

Given this sobering outcome with respect to the non-aggregated evaluation, we decided to ignore the clusters, and investigated whether parallels between the following cognitive principles could be observed: basic, contra, searchAlt and generalization. Table 5 provides an overview of the predicted conclusions according to these principles.[10] The predictions of the principles naturally raises a few questions on the plausibility of the underlying assumptions: (i) Do participants use the same reasoning strategy, i.e. can their conclusions be explained by the same principles throughout the whole reasoning task, or do they switch strategies within one reasoning task? (ii) Is the list of principles in Section 3 complete, i.e. can the conclusions of all reasoners be covered by these principles? and (iii) Do participants learn, i.e. do they improve their logical ability (with respect to CL) by doing the same reasoning task again? We address these questions in the following.

---

[10] Note that any advanced principle also includes the logic program clauses specified for **basic**.

4.2 Individual Reasoners

A wide amount of cognitive theories for syllogistic reasoning have previously only be evaluated with respect to the aggregated data of psychological experiments. Usually, as assessment, typical statistical evaluation criteria with respect to the experimental data have been used (cf. root-mean-square error, Bayesian information criterion (Schwarz, 1978), Akaike information criterion (Akaike, 1974)). These measures are helpful as a general guideline, however, their major drawback is that they only compute the quantitative overall performance. In this section, we carry out a qualitative assessment of the suggested cognitive principles with respect to the individual differences among the participants.

4.2.1 mReasoner

Johnson-Laird and Khemlani (2016) proposed an approach to model individual reasoning patterns of participants, and showed that different participants seem to derive different conclusions. Their approach is built on top of the mReasoner (Khemlani and Johnson-Laird, 2013), an extension of the mental model theory, first introduced by Johnson-Laird (1983). The reasoning procedure depends on the construction and manipulation of mental models, based on 3 fundamental principles: (i) mental models represent possibilities, (ii) mental models are iconic (principle of iconicity), and (iii) reasoning procedure consists of two interacting processes, intuition and deliberation (principle of dual processes). The procedure of the mReasoner consists of two steps: (1) The construction of a small set of models that represents the terms of the premises, and (2) the call of a deliberative component to search for counterexamples to conclusions. Johnson-Laird and Khemlani specified four parameters for the construction of a mental model and evaluated their predictions based on 2 datasets, week 1 and week 2, for 20 participants. The parameters for the construction of the mental models are (i) size of a mental model, (ii) probability of drawing conclusions only from the most common mental models, or from the complete set of models, (iii) probability to search for counter examples, and (iv) probability to weaken the conclusion. The clustering with mReasoner was based on a particular clustering algorithm to determine the optimal number of clusters for their dataset and then the mReaoner simulated the clusters by choosing appropriate parameter settings. Johnson-Laird and Khemlani (2016) identified three different reasoning clusters: The intuitive cluster, the intermediate cluster and the deliberative cluster. The approach is novel as until now no other cognitive theory tried to rigorously account for individual differences in human syllogistic reasoning. However, their approach could be optimized in the sense that the authors did not distinguish between the participants' responses in week 1 and week 2, but merged their responses. This might give a distorted picture, as participants could have learned from week 1 to week 2. We will address this issue in the following.

Table 5: Predictions according to the cognitive principles as specified in Section 3.

	Aac	Eac	Iac	Oac	Aca	Eca	Ica	Oca	NVC
AA1	basic								
AA2					basic				
AA3									basic
AA4	basic				basic				
AI1		abduction					abduction		basic
AI2		basic					basic		
AI3		abduction					abduction		basic
AI4		basic					basic		
AE1	basic								
AE2	contraposition					basic			
AE3	contraposition					contraposition			basic
AE4	basic					contraposition			
AO1				abduction					basic
AO2								basic	
AO3								contraposition	basic
AO4				basic					
IA1		basic					basic		
IA2		abduction			generalization		abduction		basic
IA3		abduction					abduction		basic
IA4		basic					basic		
II1		generalization					generalization		basic
II2		generalization					generalization		basic
II3		generalization					generalization		basic
II4		generalization					generalization		basic
IE1			basic						
IE2			contraposition		generalization				basic
IE3			contraposition		generalization				basic
IE4			basic						
IO1			generalization						basic
IO2			generalization						basic
IO3			generalization						basic
IO4			generalization						basic
EA1	basic								
EA2	contraposition					basic			
EA3	contraposition					contraposition			basic
EA4	contraposition					basic			
EI1	generalization								basic
EI2								basic	
EI3	generalization							contraposition	basic
EI4								basic	
EE1									basic
EE2									basic
EE3									basic
EE4									basic
EO1									basic
EO2									basic
EO3									basic
EO4									basic
OA1				basic					
OA2								abduction	basic
OA3				contraposition					basic
OA4								basic	
OI1				generalization					basic
OI2								generalization	basic
OI3				generalization					basic
OI4								generalization	basic
OE1									basic
OE2									basic
OE3									basic
OE4									basic
OO1									basic
OO2									basic
OO3									basic
OO4									basic

4.2.2 Characterization through Principles

Table 6: Overview of the individual participants' responses with respect to the specified criteria in week 1 and week 2. ↑, ↓, and ↔ denote above, below, and average, respectively.

Week	Part.	NVC	CL	basic	contraposition	abduction	generalization	Changes	Cluster	Cluster in (Johnson-Laird and Khemlani, 2016)
1	0	↓	↓	↓	↓	↓	↑	↑	INTUITIVE	INTUITIVE
2	0	↓	↔	↔	↑↑	↓	↓			
1	1	↓	↓	↓	↑	↑	↑	↔	INTUITIVE	INTER
2	1	↑	↑	↑	↔	↑	↓			
1	2	↓	↓↓	↓↓	↓	↑	↑↑	↑	INTUITIVE	INTUITIVE
2	2	↓	↓	↔	↓	↑	↑↑			
1	3	↓	↓↓	↓↓	↔	↑	↔	↑	INTUITIVE	INTUITIVE
2	3	↓	↓	↓	↓↓	↑	↔			
1	4	↑	↑	↑	↑↑	↓	↓↓	↓	DEL	DEL
2	4	↑	↑	↑	↑↑	↓	↓↓			
1	5	↑	↑	↑	↑	↓	↓	↓	DEL	DEL
2	5	↑	↑	↑↑	↑↑	↓	↓↓			
1	6	↔	↔	↔	↓	↔	↑↑	↑	INTER	INTER
2	6	↔	↔	↓	↑	↑↑	↑			
1	7	↑	↑	↑	↑	↓	↓	↓	DEL	DEL
2	7	↑	↑	↑	↔	↓	↓↓			
1	8	↑	↑	↑	↑	↑	↓	↓	DEL	DEL
2	8	↑	↑	↑	↑↑	↑	↓↓			
1	9	↔	↔	↔	↔	↑	↑	↑	INTER	INTER
2	9	↔	↑	↑	↔	↓	↓↓			
1	10	↑	↑	↑	↑	↓	↓↓	↓	DEL	DEL
2	10	↑	↑	↑↑	↑	↓	↓↓			
1	11	↓	↔	↑	↑	↑	↑	↓	INTER	INTUITIVE
2	11	↓	↓	↓	↓	↑	↑			
1	12	↔	↔	↔	↔	↓	↔	↔	INTER	INTER
2	12	↔	↔	↔	↓	↑	↔			
1	13	↔	↓	↔	↑	↓	↑	↑	INTER	INTER
2	13	↑	↓	↔	↓	↑	↓			
1	14	↓	↓	↑	↓	↑↑	↑↑	↓	INTER	INTER
2	14	↑	↔	↑↑	↓	↑	↑			
1	15	↓	↓	↑	↓↓	↓	↑	↑↑	INTUITIVE	INTUITIVE
2	15	↓	↓	↓	↓	↑↑	↑↑			
1	16	↑	↓	↔	↓	↑	↓	↑	INTER	INTER
2	16	↓	↓	↔	↑	↑	↓			
1	17	↓	↓	↓	↓	↑↑	↑↑	↑	INTUITIVE	INTUITIVE
2	17	↔	↓	↓	↑↑	↑	↑↑			
1	18	↑	↑	↑	↑	↓	↓↓	↓	DEL	DEL
2	18	↑	↑	↑↑	↓	↓	↓↓			
1	19	↔	↓	↑	↓	↓	↓	↑	INTUITIVE	INTER
2	19	↔	↔	↓	↓	↓	↓			

Taking the non-aggregated data from Johnson-Laird and Khemlani (2016), we investigated how well the participants' responses could be explained through the cognitive principles, and whether their individual responses would be consistent with respect to a certain set of principles. Recall, Johnson-Laird and Khemlani's dataset consisted of responses from 20 participants, with each two entries, as the task was carried out twice, with a break of one week.

We observed the following patterns: On average, the basic principles explained the results about 10 % better than CL. Generally, the participants who were strong in CL/ contraposition, drew fewer conclusions according to abduction/ generalization, and vice versa. The changes within participant's responses from one to the other week was high, with 52/64 on average. From week 1 to week 2, an increase on conclusions corresponding to CL (19/20, 6 %), basic (18/20, 6 %), contraposition (15/20, 5 %) and a decrease on conclusions corresponding to generalization (15/20, 8 %) and no change on conclusions corresponding to abduction (10/20, 0 %) could be observed.

Table 6 provides an overview of the performance of each participant (column 2), for week 1 and week 2 (column 1) with respect to CL (column 4) and the cognitive principles (column 5, 6, 7, and 8). NVC (column 3) denotes how frequently participants responded *no valid conclusion* and Changes (column 9) denotes how frequently participants changed their responses from week 1 to week 2. The symbols ↑, ↓, and ↔ denote above, below, and average, respectively.

We identified three types of reasoners, characterizing them according to the terms already used in (Johnson-Laird and Khemlani, 2016): The last two columns show the classification of the participants as deliberative, intuitive or intermediate. Our classification only differs with respect to Johnson-Laird and Khemlani's classification for the intermediate and intuitive reasoners with respect to three participants (1, 11 and 19). A possible explanation for this difference could be that they did not take into account the participants' changes between week 1 and week 2. Two of the three types of reasoners can be characterized through a set of cognitive principles, but for one group of participants, no distinguishable pattern could be identified:

Deliberative reasoners (6 out of 20 participants) are characterized by the participants whose responses corresponded to the predictions made by basic/ contraposition. Almost none of their conclusions corresponded to predictions made by abduction/ generalization. Theses participants did change their responses from week 1 to week 2 lower than average and concluded higher than average NVC. Generally, their responses improved from week 1 to week 2 with respect to CL.

Intuitive reasoners (7 out of 20 participants) are characterized by the participants whose responses corresponded to the predictions given by abduction/ generalization. Almost none of their conclusions corresponded to the predictions made by contraposition (except for 2 participants in week 2). They showed a high change within their own responses from week 1 to week 2 and concluded lower than average NVC. Generally, their responses improved from week 1 to week 2 with respect to CL, but they never performed *as well as* the participants in the deliberative cluster.

Intermediate reasoners (7 out of 20 participants) are characterized by the participants whose responses did not corresponded to any particular set of principles. These participants had a high change within their own responses from week 1 to week 2 and no particular pattern with respect to NVC conclusions. Generally they were not improving their responses from week 1 to week 2 with respect to CL, but normally *compensated* for responses valid in CL in some cases (that corresponded to contraposition) with responses not valid in CL in the other cases (that correspond to abduction/ generalization).

5 Conclusions and Future Directions

We have investigated the cognitive adequacy of clustering by principles with respect to the individual response patterns reported by Johnson-Laird and Khemlani (2016). Their approach is the only one we are aware of that has evaluated predictions on human syllogistic reasoning with respect to the non-aggregated data. Here, we additionally took into account the differences between the participants' responses in week 1 and week 2. Our results are two-fold: First, the reasoning clusters as specified in (Dietz Saldanha et al, 2017), could not be validated with respect to a dataset on individual reasoners. However, we could show that the response patterns of the individual participants corresponded to new sets of cognitive principles. In particular, some principles seemed to exclude each other throughout the experiment (contraposition and abduction), whereas others often appeared together (basic with contraposition and abduction with generalization). Additionally, the provided data allowed us to observe other patterns (high/ low frequency of NVC and high/ low frequency of changing responses between week 1 and week 2). Interestingly, there was one group of participants for which we could not identify any pattern (intermediate).

For the future, we intend to investigate whether the sequential application of the cognitive principles can give us more insights on the participants' reasoning pattern. In particular, our hypothesis is that these principles might correspond to some of the reasoning steps within the insight model by Johnson-Laird and Wason (1970) for the selection task (Wason, 1968). Similarly, a meta-analysis of the individual response patterns for the selection task has been provided by Ragni, Kola and Johnson-Laird (2017). This in turn might give us a new starting point to investigate whether the cognitive principles specified here can characterize participants' responses in other reasoning episodes. Furthermore, it is quite likely that the set of cognitive principles presented in Section 3 is not complete, and thus other principles will be identified in the future.

References

Akaike H (1974) A new look at the statistical model identification. IEEE Transactions on Automatic Control 19(6):716–723

Barnes J (ed) (1984) The Complete Works of Aristotle (Vol. 1). Princeton University Press, Princeton, NJ

Byrne RM (1989) Suppressing valid inferences with conditionals. Cognition 31:61–83

Byrne RM, Johnson-Laird PN(1989) Spatial reasoning. Journal of Memory and Language 28(5):564–575

Chater N, Oaksford M (1999) The probability heuristics model of syllogistic reasoning. Cognitive Psychology 38(2):191 – 258

Costa A, Dietz Saldanha EA, Hölldobler S (2017a) Monadic reasoning using weak completion semantics. In: Hölldobler S, Malikov A, Wernhard C (eds) Proceedings of the Young Scientist's Second International Workshop on Trends in Information Processing (YSIP 2), CEUR-WS.org, CEUR Workshop Proceedings

Costa A, Dietz Saldanha EA, Hölldobler S, Ragni M (2017b) A computational logic approach to human syllogistic reasoning. In: Gunzelmann G, Howes A, Tenbrink T, Davelaar E (eds) Proceedings of the 39th Annual Conference of the Cognitive Science Society, (CogSci 2017), Cognitive Science Society, pp 883–888

Dietz EA, Hölldobler S, Wernhard C (2014) Modeling the suppression task under weak completion and well-founded semantics. Journal of Applied Non-Classsical Logics 24(1-2):61–85

Dietz Saldanha EA, Hölldobler S, Mörbitz R (2017) The syllogistic reasoning task: Reasoning principles and heuristic strategies in modeling human clusters. In: Seipel D, Hanus M, Abreu S (eds) Proceedings of the 21st International Conference on Applications of Declarative Programming and Knowledge Management (INAP 2017), Springer, Lecture Notes in Artificial Intelligence, pp 139–154

Dietz Saldanha EA, Hölldobler S, Pereira LM (2017) Contextual reasoning: Usually birds can abductively fly. In: Proceedings of 14th International Conference on Logic Programming and Nonmonotonic Reasoning (LPNMR), Springer International Publishing, pp 64–77

Dietz Saldanha EA, Hölldobler S, Philipp T (2017) Contextual abduction and its complexity issues. In: Proceedings of the 4th International Workshop on Defeasible and Ampliative Reasoning (DARe) co-located with the 14th International Conference on Logic Programming and Nonmonotonic Reasoning (LPNMR), CEUR-WS.org, CEUR Workshop Proceedings, vol 1872, pp 58–70

Gebser M, Kaminski R, Kaufmann B, Schaub T (2012) Answer Set Solving in Practice. Synthesis Lectures on Artificial Intelligence and Machine Learning, Morgan & Claypool Publishers

Grice HP (1975) Logic and conversation. In: Cole P, Morgan JL (eds) Syntax and Semantics: Vol. 3: Speech Acts, Academic Press, New York, pp 41–58, reprinted as ch.2 of Grice 1989, pages 22–40

Hölldobler S (2015) Weak completion semantics and its applications in human reasoning. In: Furbach U, Schon C (eds) CEUR WS proc. on Bridging the Gap between Human and Automated Reasoning, pp 2–16

Hölldobler S, Kencana Ramli CDP (2009) Logic programs under three-valued łukasiewicz semantics. In: Hill PM, Warren DS (eds) Proc. of 25th Int. Conf. on Logic Programming, Springer Berlin Heidelberg, LNCS, vol 5649, pp 464–478

Hölldobler S, Kencana Ramli CDP (2009) Logics and networks for human reasoning. In: Alippi C, Polycarpou MM, Panayiotou CG, Ellinas G (eds) International Conference on Artificial Neural Networks, (ICANN 2009), Part II, Springer, Heidelberg, Lecture Notes in Computer Science, vol 5769, pp 85–94

Johnson-Laird P, Wason PN (1970) A theoretical analysis of insight into a reasoning task. Cognitive Psychology 1(2):134 – 148

Johnson-Laird PN (1983) Mental models: towards a cognitive science of language, inference, and consciousness. Harvard University Press, Cambridge, MA

Johnson-Laird PN, Khemlani SS (2016) How people differ in syllogistic reasoning. In: Papafragou A, Grodner D, Mirman D, Trueswell J (eds) Proceedings of the 38th

Annual Conference of the Cognitive Science Society, (CogSci 2016), Austin, TX: Cognitive Science Society, pp 2165–2170

Khemlani S, Johnson-Laird PN (2012) Theories of the syllogism: A meta-analysis. Psychological Bulletin pp 427–457

Khemlani S, Johnson-Laird PN (2013) The processes of inference. Argument & Computation 4:1–20

Knauff M, Rauh R, Renz J (1997) A cognitive assessment of topological spatial relations: Results from an empirical investigation. In: Proceedings of the third International Conference on Spatial Information Theory, (COSIT 1997), Springer, Heidelberg, Lecture Notes in Computer Science, vol 1329, pp 193–206

Kotarbiński T (1913) Zagadnienie istnienia przyszłości (the problem of existence of the future). Przeglad Filozoficzny VI.1

Łukasiewicz J (1920) O logice trójwartościowej. Ruch Filozoficzny 5:169–171, english translation: On three-valued logic. In: Łukasiewicz J. and Borkowski L. (ed.). (1990). *Selected Works*, Amsterdam: North Holland, pages 87–88.

Malinowski G (1993) Many-Valued Logics. Oxford University Press

O'Brien D, D S Braine M, Yang Y (1994) Propositional reasoning by mental models? simple to refute in principle and in practice 101:711–24

Parry W, Hacker E (1991) Aristotelian Logic. G - Reference,Information and Interdisciplinary Subjects Series, State University of New York Press

Polk TA, Newell A (1995) Deduction as verbal reasoning. Psychological Review 102(3):533–566

Ragni M, Stolzenburg F (2015) Higher-level cognition and computation: A survey 29:247–253

Ragni M, Kola I, Johnson-Laird P (2017) The wason selection task: A meta-analysis. In: Gunzelmann G, Howes A, Tenbrink T, Davelaar E (eds) Proc. of 39th Annual Conference of the Cognitive Science Society, Cognitive Science Society, pp 980–985

Riefer DM, Batchelder WH (1988) Multinomial modeling and the measurement of cognitive processes. Psychological Review 95(3):318–339

Rips LJ (1994) The psychology of proof: Deductive reasoning in human thinking. The MIT Press, Cambridge, MA

Schwarz G (1978) Estimating the dimension of a model. Ann Statist 6(2):461–464, DOI 10.1214/aos/1176344136

Stenning K, van Lambalgen M (2008) Human Reasoning and Cognitive Science. A Bradford Book, MIT Press, Cambridge, MA

Strube G (1992) The role of cognitive science in knowledge engineering. In: Schmalhofer F, Strube G, Wetter T (eds) Contemporary Knowledge Engineering and Cognition, First Joint Workshop, Lecture Notes in Computer Science, vol 622, Springer, Heidelberg, pp 159–174

Wason PC (1968) Reasoning about a rule. Quarterly Journal of Experimental Psychology 20(3):273–281

Wetherick NE, Gilhooly KJ (1995) 'atmosphere', matching, and logic in syllogistic reasoning. Current Psychology 14(3):169–178

Woodworth RS, Sells SB (1935) An atmosphere effect in formal syllogistic reasoning. Journal of Experimental Psychology 18(4):451–60

Acceptable propositional normal logic programs checking procedure implementation

Aleksandra Czyż, Kinga Ordecka, Andrzej Gajda

Abstract Acceptable normal logic programs is a class of normal logic programs for which the immediate consequence operator always has a fixpoint. However, the definition was made specifically for Prolog implementation of logic programs, therefore it is not optimal concerning other implementations of logic programs. In this work we propose solutions for three main issues that increase the computational cost of checking, if a given logic program is acceptable. We also propose an implementation of the procedure written in Haskell programming language.

Key words: logic programs, acceptable logic programs, immediate consequence operator

---

Aleksandra Czyż
Reasoning Research Group
Faculty of Psychology and Cognitive Science
Adam Mickiewicz University, Poznań, Poland
e-mail: aleksandra.m.czyz@gmail.com

Kinga Ordecka
Reasoning Research Group
Faculty of Psychology and Cognitive Science
Adam Mickiewicz University, Poznań, Poland
e-mail: kingaordecka@gmail.com

Andrzej Gajda
Department of Logic and Cognitive Science
Faculty of Psychology and Cognitive Science
Adam Mickiewicz University, Poznań, Poland
e-mail: andrzej.gajda@amu.edu.pl

1 Introduction

The neural-symbolic integration movement which combine the study of logic with connectionism allow to create intelligent systems that benefit from both approaches (Hilario, 2013; Garcez et al, 2002). In this article we want to present part of the implementation of a neural-symbolic system that is used for modelling abductive reasoning (Gajda et al, 2016). In this approach the Connectionist Inductive Learning and Logic Programming system (C-IL^2P) (Garcez et al, 2002) is used, which integrates logic programs (Lloyd, 1993) and artificial neural networks by allowing to translate symbolic knowledge in the form of a logic program into a neural network, and inversely, to translate a neural network into a set of logical rules. Since neural networks that result from the translation of logic programs model the way the immediate consequence operator works for those logic programs, the process requires to check if every logic program (before translation into a neural network and after the translation from the neural network) has the property of being acceptable, i.e. if the immediate consequence operator has a fixpiont, and what follows, if the neural network can stabilise itself. This means that this particular procedure can be potentially executed thousands of times during our research. Therefore, it is important for the implementation of the algorithm that allows to check if a given logic program is acceptable to be very efficient. However, there are some issues concerning the procedure of acceptability checking of a given logic program and the goal of this article is to present solutions that we have applied in our implementation[1].

The article is structured as follows: in the two next sections we introduce theory of logic programs and definitions that concern acceptability of logic programs. At the end of the third section we address problems that arises from the definition of an acceptable logic program and in the following section we present our solutions for those problems. We end with a brief summary.

2 Normal logic programs

The language $\mathcal{L}$ in which we define normal logic programs consists of the following elements: $\{a_0, a_1, a_2, \ldots\}$ — an infinite, countable set of propositional variables, $\leftarrow$, $\sim$ — primitive connectives and a comma. Additionally, we refer to atoms (a_i) or their negations ($\sim a_i$) as *literals* (sometimes we will use A and L as a meta atom and a meta literal variable, respectively, to avoid double indexation).

Horn clauses (or just *clauses*) denoted by h_i are structures of the following form:

$$a_j \leftarrow a_k, \ldots, a_l, \sim a_m, \ldots, \sim a_n$$

where atom a_j is the *head* of Horn clause h_i (denoted by $head(h_i)$), atoms $a_k, \ldots, a_l$ form the *positive body* of Horn clause h_i (denoted by $body_p(h_i)$), atoms that are preceded by the negation, i.e. atoms $a_m, \ldots, a_n$, form the *negative body* of Horn clause

---
[1] The code is available at https://github.com/flaizdnag/lph.

h_i (denoted by $body_n(h_i)$), and finally, the sum of both sets of atoms, the positive and negative body, form the *atomic body* of Horn clause h_i (denoted by $body_a(h_i)$).

Propositional normal logic programs are defined as finite and non-empty sets of Horn clauses and are denoted by $\mathcal{P}_i$. The set of all atoms that occur in a logic program $\mathcal{P}_i$ is called the *Herbrand base* for $\mathcal{P}_i$ and is denoted by $B_{\mathcal{P}_i}$.

Now we are going to define a semantics for normal logic programs in terms of Herbrand interpretations. A Herbrand interpretation $I_{\mathcal{P}_i}$ for a logic program $\mathcal{P}_i$ is a set of atoms that occur in $\mathcal{P}_i$ and are mapped to *true*. If interpretation $I_{\mathcal{P}_i}$ contains an atom a_j, then the negation of that atom, i.e. $\sim a_j$, is mapped to *false*. A Horn clause h_k is mapped to *true* w.r.t. an interpretation $I_{\mathcal{P}_i}$ iff its head is mapped to *true* or one of the literals from its body is mapped to *false*. An interpretation $I_{\mathcal{P}_i}$ is a model for a logic program $\mathcal{P}_i$ iff all clauses from $\mathcal{P}_i$ are mapped to *true* w.r.t. $I_{\mathcal{P}_i}$. What can be observed is that the power set of the Herbrand base $B_{\mathcal{P}_i}$ (denoted by $\wp(B_{\mathcal{P}_i})$) contains all possible Herbrand interpretations for a logic program $\mathcal{P}_i$.

Van Emden and Kowalski [1976] defined the immediate consequence operator, which is a mapping that takes a Herbrand interpretation as an argument and returns heads of all those Horn clauses that positive bodies are subsets of this interpretation, while none of atoms that belong to the negative body is an element of the interpretation. By means of the immediate consequence operator a Herbrand model for a logic program can be found.

Definition 2.1. *Let $\mathcal{P}_i$ be a normal logic program and $I_{\mathcal{P}_i}$ a Herbrand interpretation. The mapping $T_{\mathcal{P}_i} : \wp(B_{\mathcal{P}_i}) \longrightarrow \wp(B_{\mathcal{P}_i})$ is defined as follows:*

$$T_{\mathcal{P}_i}(I_{\mathcal{P}_i}) =_{df} \{head(h_j) \mid h_j \in \mathcal{P}_i, body_p(h_j) \subseteq I_{\mathcal{P}_i},$$
$$body_n(h_j) \cap I_{\mathcal{P}_i} = \varnothing\}$$

Searching for the model by means of the $T_{\mathcal{P}_i}$ operator is done by applying the operator to the result of the previous application, where the starting point is the empty set (empty interpretation) and the end point is the fixpoint. For example, for logic the program:

$$\mathcal{P}_1 = \{ \quad a_2 \leftarrow a_1, \sim a_3$$
$$a_4 \leftarrow a_1 \quad \quad (1)$$
$$a_3 \leftarrow a_4 \quad \}$$

the search for the model looks as follows:

$$T_{\mathcal{P}_1} \uparrow 0 = \varnothing$$
$$T_{\mathcal{P}_1} \uparrow 1 = T_{\mathcal{P}_1}(\varnothing) = \varnothing$$

However, the immediate consequence operator $T_{\mathcal{P}_i}$ for normal logic programs is not a monotonic mapping, therefore, there are logic programs for which $T_{\mathcal{P}_i}$ operator does not have a fixpoint, what results in its "inability to stop". Let us consider the following example of a logic program $\mathcal{P}_2$:

$$\mathcal{P}_2 = \{a_1 \leftarrow \sim a_1\} \qquad (2)$$

and the way the immediate consequence operator works for this logic program:

$$T_{\mathcal{P}_2} \uparrow 0 = \varnothing$$
$$T_{\mathcal{P}_2} \uparrow 1 = \{a_1\}$$
$$T_{\mathcal{P}_2} \uparrow 2 = \varnothing$$
$$\vdots$$

The neural network that results from the translation of such a logic program cannot stabilise itself, what reflects the situation when the immediate consequence operator does not have a fixpoint.

3 Acceptable normal logic programs

Acceptable logic programs are logic programs for which the immediate consequence operator always reaches a fixpoint. The rough idea is the following: let us assume that we can travel from an atom a_1 to an atom a_2, if a_1 is the head of a clause (let us call it h_1) and a_2 (or $\sim a_2$) is an element of the body of this clause (i.e. h_1), or a_2 (or $\sim a_2$) is an element of the body of a clause (let us call it h_2) whose head (or negated head) is an element of the body of the previous clause (i.e. h_1), or a_2 (or $\sim a_2$) is an element of the body of the of a clause (let us call it h_3) whose head (or negated head) is an element of the body of a clause (h_2) whose head (or negated head) is an element of the body of the initial clause h_1, or etc. In example 1 we can travel from atom a_2 to atom a_4 "through" atom a_3. Now, a logic program is acceptable if we cannot travel from atom to its negated version (the simplest example is $\mathcal{P}_2$ from example 2).

In order to check if a given logic program is acceptable we have to introduce additional technical notions. The following definitions are based on the article by Apt and Pedreschi (Apt and Pedreschi, 1993, p. 128–129).

There are three main components that the definition of an acceptable logic program is based on. The first one is a logic program $\mathcal{P}_i^-$, which is a subset of clauses from the initial logic program $\mathcal{P}_i$. In order to establish which clauses from $\mathcal{P}_i$ are part of the $\mathcal{P}_i^-$ we have to build a graph for the logic program $\mathcal{P}_i$ and establish connections between atoms that occur negatively in $\mathcal{P}_i$ and others.

Definition 3.1. *Let $\mathcal{P}_i$ be a logic program and $B_{\mathcal{P}_i}$ the Herbrand base of $\mathcal{P}_i$. $G_{\mathcal{P}_i}$ is a directed graph for $\mathcal{P}_i$, where the set of nodes N is equal to $B_{\mathcal{P}_i}$ and for every $h_j \in \mathcal{P}_i$:*

- *if $a_m = head(h_j)$, then for every $a_n \in body(h_j)$: $edge(a_m, a_n) \in G_{\mathcal{P}_i}$,*
- *if $edge(a_m, a_n) \in G_{\mathcal{P}_i}$, then $path(a_m, a_n) \in G_{\mathcal{P}_i}$,*
- *if $edge(a_m, a_n) \in G_{\mathcal{P}_i}$ and $path(a_n, a_o) \in G_{\mathcal{P}_i}$, then $path(a_m, a_o) \in G_{\mathcal{P}_i}$.*

The graph $G_{\mathcal{P}_1}$ for the logic program $\mathcal{P}_1$ from example 1 is given in Figure 1.

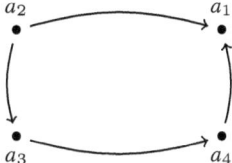

Fig. 1: Directed graph $G_{\mathcal{P}_1}$ for the logic program $\mathcal{P}_1$.

The following two definitions cover the set of atoms that occur negatively in a logic program $\mathcal{P}_i$ and the set of atoms that are connected to those negatively occurring atoms in the graph $G_{\mathcal{P}_i}$ for the logic program $\mathcal{P}_i$.

Definition 3.2. *Let $\mathcal{P}_i$ be a logic program. The set of all atoms that occur negatively in $\mathcal{P}_i$ is defined in the following way:*

$$Neg_{\mathcal{P}_i} =_{df} \{a_j \mid a_j \in body_n(h_k), h_k \in \mathcal{P}_i\}$$

Definition 3.3. *Let $\mathcal{P}_i$ be a logic program and $Neg_{\mathcal{P}_i}$ the set of all atoms that occur negatively in $\mathcal{P}_i$. By $Neg^*_{\mathcal{P}_i}$ we define the following set:*

$$Neg^*_{\mathcal{P}_i} =_{df} \{a_j \in B_{\mathcal{P}} \mid path(a_k, a_j) \in G_{\mathcal{P}_i}, a_k \in Neg_{\mathcal{P}_i}\}$$

A logic program $\mathcal{P}_i^-$ is a set of clauses from $\mathcal{P}_i$ whose heads are elements of $Neg^*_{\mathcal{P}_i}$.

Definition 3.4. *Let $\mathcal{P}_i$ be a logic program. We define logic program $\mathcal{P}_i^-$ in the following way:*

$$\mathcal{P}_i^- =_{df} \{h_j \in \mathcal{P}_i \mid head(h_j) \in Neg^*_{\mathcal{P}_i}\}$$

Looking back at our logic program from example 1 for which we created a graph $G_{\mathcal{P}_1}$ (see Fig. 1) we have the following:

$$\begin{aligned}
\mathcal{P}_1 &= \{a_2 \leftarrow a_1, \sim a_3 \;;\; a_4 \leftarrow a_1 \;;\; a_3 \leftarrow a_4\} \\
Neg_{\mathcal{P}_1} &= \{a_3\} \\
Neg^*_{\mathcal{P}_1} &= \{a_4\} \\
\mathcal{P}_1^- &= \{a_4 \leftarrow a_1\}
\end{aligned} \qquad (3)$$

The second component is the Clark's completion of a logic program, i.e. the interpretation of a logic program in classical propositional logic (CPL). The language $\mathcal{L}_{CPL}$ of CPL is defined as usual, with the addition of the top element $\top$. Similarly, the set of formulas of language $\mathcal{L}_{CPL}$ is defined in a standard way. The definition of Clark's completion (or just *completion*) of a logic program is based on the notion of a *df* of an atom a_j in the logic program $\mathcal{P}_i$, which is the set of all clauses from $\mathcal{P}_i$ that have a_j as the head, and a *completed definition* of an atom a_j in the logic program $\mathcal{P}_i$, which is an expression of the $\mathcal{L}_{CPL}$ language resulting from the transformation of the definition of a_j in $\mathcal{P}_i$.

Definition 3.5. *Let a_j be an atom and $\mathcal{P}_i$ a logic program. The* df *of a_j in $\mathcal{P}_i$ is the following set:*

$$d(a_j, \mathcal{P}_i) = \{h_k \in \mathcal{P}_i \mid a_j = head(h_k)\}$$

Definition 3.6. *Let a_j be an atom, $\mathcal{P}_i$ a logic program and $d(a_j, \mathcal{P}_i)$ a definition of a_j in $\mathcal{P}_i$. Let us assume that the definition of a_j in $\mathcal{P}_i$ is of the following form:*

$$d(a_j, \mathcal{P}_i) = \{a_j \leftarrow B_1, a_j \leftarrow B_2, \ldots, a_j \leftarrow B_n\}$$

where B_k ($1 \leq k \leq n$) is the body of k-th Horn clause from $d(a_j, \mathcal{P}_i)$. The completed definition *of a_j in $\mathcal{P}_i$ denoted by $cd(a_j, \mathcal{P}_i)$ is the following expression:*

$$a_j \rightarrow (B'_1 \vee B'_2 \vee \ldots \vee B'_n)$$

where B'_k ($1 \leq k \leq n$) is obtained from B_k by replacing the commas separating literals by $\wedge$ and all occurrences of default negation $\sim$ by classical negation $\neg$. If it is the case that B_k is empty, i.e. a_j is a fact, then the top element $\top$ is inserted instead of B'_k.

Clark's completion of a logic program $\mathcal{P}_i$ consists of clauses from $\mathcal{P}_i$ turned into implications from $\mathcal{L}_{\text{CPL}}$, completed definitions of all atoms that are heads of clauses in $\mathcal{P}_i$ and classically negated atoms that have empty definitions in $\mathcal{P}_i$.

Definition 3.7. *Let $\mathcal{P}_i$ be a logic program. The* Clark's completion *of $\mathcal{P}_i$ denoted by $comp(\mathcal{P}_i)$ is a set of clauses of $\mathcal{L}_{\text{CPL}}$ containing the following elements:*

- *clauses from $\mathcal{P}_i$ with default negation ($\sim$) replaced by the classical negation ($\neg$), commas in the bodies replaced by conjunction ($\wedge$) and arrow ($\leftarrow$) treated as a material implication,*
- *completed definitions of atoms defined in $\mathcal{P}_i$,*
- *expressions of the form: $\neg a_j$, for every atom $a_j \in B_{\mathcal{P}_i}$ not defined in $\mathcal{P}_i$.*

For example, the completed version of the logic program $\mathcal{P}_1$ from the example 1 looks as follows:

$$\begin{aligned} comp(\mathcal{P}_1) = \{&a_2 \leftrightarrow a_1 \wedge \neg a_3 \\ &a_4 \leftrightarrow a_1 \\ &a_3 \leftrightarrow a_4 \\ &\neg a_1 \quad \} \end{aligned} \qquad (4)$$

Clauses from a logic program interpreted as implications along with completed definitions of atoms that are heads from that logic program together form equivalences, therefore in the example 4 we have shown version of the completion that already has equivalences instead of implications going in both directions.

Finally, we define a *level mapping* function, which assigns to atoms from the Herbrand base $B_{\mathcal{P}_i}$ of a logic program $\mathcal{P}_i$ arbitrary natural numbers.

Definition 3.8. *Let $\mathcal{P}_i$ be a logic program and $B_{\mathcal{P}_i}$ a Herbrand base of $\mathcal{P}_i$. A level mapping is a function $|\cdot|: B_{\mathcal{P}_i} \longrightarrow \mathbb{N}$. $|a_j|$ is the level of a_j and $|a_j| = |\sim a_j|$.*

In the following definition of an acceptable logic program we have restricted interpretations to Herbrand interpretations which is sufficient in the case of propositional version of logic programs.

Definition 3.9. *Let $\mathcal{P}_i$ be a logic program, $|\cdot|$ a level mapping for $\mathcal{P}_i$, and $I_{\mathcal{P}_i}$ a model for $\mathcal{P}_i$ whose restriction to the atoms in $Neg^*_{\mathcal{P}_i}$ is a model for the completion of $\mathcal{P}_i^-$. Logic program $\mathcal{P}_i$ is called* acceptable *w.r.t. $|\cdot|$ and $I_{\mathcal{P}_i}$ if, for every clause $A \leftarrow L_1, \ldots, L_n$ in $\mathcal{P}_i$:*

$$|A| > |L_j| \text{ for } j \in [1, \bar{n}]$$

where

$$\bar{n} = \min(\{n\} \cup \{j \in [1, n] \mid I_{\mathcal{P}_i} \nvDash L_j\})$$

$\mathcal{P}_i$ is called acceptable *if it is acceptable w.r.t. some level mapping and a model for $\mathcal{P}_i$.*

There are three main issues with the definition of acceptable logic programs that we addressed in our implementation. The first one concerns the arbitrariness of the model for the logic program $\mathcal{P}_i$, which restricted to the atoms in $Neg^*_{\mathcal{P}_i}$ must be a model for the completion of $\mathcal{P}_i^-$. Let us consider the logic program $\mathcal{P}_1$ and the following interpretation that is his model:

$$I_{\mathcal{P}_1} = \{a_3, a_4\}$$

Recalling that (see (3))

$$Neg^*_{\mathcal{P}_1} = \{a_4\}$$
$$\mathcal{P}_1^- = \{a_4 \leftarrow a_1\}$$

we can see that the model $I_{\mathcal{P}_1}$ for $\mathcal{P}_1$ does not meet our condition, because the completion of $\mathcal{P}_1^-$ is the following set (recall that an interpretation contains atoms mapped to *true*, therefore atoms that are not in the interpretation are mapped to *false*):

$$\text{comp}(\mathcal{P}_1^-) = \{a_4 \leftrightarrow a_1, \neg a_1\}$$

Therefore the question arises how to efficiently generate models for logic programs that also fulfil the above-mentioned criterion.

The second issue concerns possible level mappings. The definition 3.8 does not state how to assign natural numbers for atoms, which means that testing random level mappings one after another potentially accounts for a lot of calculations, especially when logic program contains a large number of atoms.

Finally, the notion of acceptability of a logic program was defined as a tool for checking if a program written in Prolog can successfully execute the computation. Because of the Prolog execution rule, which checks atoms from the body of a given Horn clause in specific order: from left to right, the placement of atoms in the body of Horn clauses can influence the result of the acceptability property checking. For example, from the two logic programs (see Garcez et al, 2002, p. 34)

$$\mathcal{P}_3 = \{a_1 \leftarrow \sim a_1, a_2\}$$
$$\mathcal{P}_4 = \{a_1 \leftarrow a_2, \sim a_1\}$$

only the second one, i.e. $\mathcal{P}_4$, is acceptable. Since the ordering of atoms in the bodies of clauses is not of importance to us, we would have to check the property for each logic program for every combination of atoms in the bodies in its clauses.

In the next section we describe our solutions to those three issues that allow us to avoid at least part of calculations in the majority of cases.

4 Implementation

The implementation of the procedure was conveyed in Haskell programming language (Hutton, 2016), which is a purely functional programming language. All of the code is available at our repository: https://github.com/flaizdnag/lph.

In order to reduce the number of generated models for a logic program $\mathcal{P}_i$ we took the following strategy:

1. Generate $\text{comp}(\mathcal{P}_i^-)$ and $Neg^*_{\mathcal{P}_i}$ for $\mathcal{P}_i$.
2. Create the power set $\wp(Neg^*_{\mathcal{P}_i})$ for the set $Neg^*_{\mathcal{P}_i}$.
3. Filter out those elements of $\wp(Neg^*_{\mathcal{P}_i})$ that are not models for $\text{comp}(\mathcal{P}_i^-)$.
4. For each element from $\wp(Neg^*_{\mathcal{P}_i})$ that remained after the filtering from the previous step generate all possible models for $\mathcal{P}_i$.

The last step, i.e. the generation of list of all possible models for a given list of atoms, is conducted by the addition of all combinations of the remaining atoms from the Herbrand base of the program $\mathcal{P}_i$ after removal of elements that belong to the list being processed. Therefore, it should be noted that in the worst case scenario the complexity is the same as in the case of testing all possible models for the program $\mathcal{P}_i$. However, this is rather a rare case, since in larger programs sets $Neg^*_{\mathcal{P}_i}$ and $\mathcal{P}_i^-$ are usually not empty.

In the case of the second problem it appears that we are able to reduce the search space of all possible level mappings by identifying those atoms, for which we do not have to change level. In general, the condition in the definition of an acceptable logic program requires that the level of the atom in the head of a clause is greater than the level of some atoms in the bodies. Therefore, we can divide atoms from a logic program into three groups: atoms that appear only in bodies of clauses (let us assume that there are n_b of them), atoms that appear only in heads of clauses (let us assume that there are n_h of them), and the rest, i.e. atoms that appear in both: bodies and heads (let us assume that there are n_r of them). Now we assign levels to atoms in the following way (n stands for the number of atom in the considered logic program):

- atoms that appear only in bodies are assigned to numbers $[1, 2, \ldots, n_b]$,
- atoms that appear only in heads are assigned to numbers $[n - n_h + 1, n - n_h + 2, \ldots, n]$,

- atoms that appear in bodies and heads are assigned to numbers $[n_b + 1, n_b + 2, \ldots, n - n_h]$.

Only the last set of atoms is "problematic", i.e. only atoms that appear in heads as well as in bodies of clauses should be tested with different level mappings. Therefore, we create all permutations of assignments in the last set and combine each permutation with the first and the second set of assignments. The result is the list of all level mappings that we have to consider in order to check all possibilities that could be "good". If there are atoms in the Herbrand base of a given logic program that do not occur as both: a head and element of a body of some clauses, then the number of level mapping we have to check is smaller than the number of all possible level mappings for that program (which is usually the case).

Finally, in order to avoid generating and checking every possible combination of atoms in the bodies of clauses from a given logic program we perform the following steps for every clause h_i:

1. Search for literals in the body of the clause h_i that do not follow from a given interpretation.
2. If there are no such literals, then the clause h_i fulfils the condition given in the definition 3.9 if for all $l_j \in body(h_i)$: $\mid head(h_i) \mid > \mid l_j \mid$.
3. Otherwise, the clause h_i fulfils the condition given in the definition 3.9 if there exists a literal l_j which belongs to the set created in the first step and $\mid head(h_i) \mid > \mid l_j \mid$.

Taking all things together we say that a logic program $\mathcal{P}_i$ is acceptable if there exists level a mapping in generated level mappings for which there exists an interpretation in generated interpretations, for which all clauses from $\mathcal{P}_i$ fulfil the condition described above in three steps.

5 Summary and conclusions

The immediate consequence operator always has a fixpoint for acceptable logic programs. However, the procedure for checking if a given logic program is acceptable that is based on the definition of an acceptable logic program is not very efficient. In order to enhance the performance of out implementation we had to address three main problems: the generation of models for our logic program that we have to check; the generation of level mappings for atoms from our logic program; the susceptibility of the procedure to the ordering of atoms in bodies of clauses in our logic program. The first two problems were addressed with methods that allow us to generate as few cases that we have to check as it is possible. The last problem was eliminated completely.

References

Apt K, Pedreschi D (1993) Reasoning about termination of pure prolog programs. Information and computation 106(1):109–157

Gajda A, Kups A, Urbański M (2016) A connectionist approach to abductive problems: employing a learning algorithm. In: Ganzha M, Maciaszek L, Paprzycki M (eds) Proceedings of the 2016 Federated Conference on Computer Science and Information Systems, IEEE, Annals of Computer Science and Information Systems, vol 8, pp 353–362

Garcez ASd, Broda K, Gabbay DM (2002) Neural-Symbolic Learning Systems: Foundations and Applications. Springer-Verlag, London

Hilario M (2013) An overview of strategies for neurosymbolic integration. In: Sun R, Alexandre F (eds) Connectionist-Symbolic Integration: From Unified to Hybrid Approaches, Psychology Press, chap 2, pp 13–36

Hutton G (2016) Programming in Haskell. Cambridge University Press

Lloyd JW (1993) Foundations of logic programming. Springer-Verlag, Berlin

Van Emden MH, Kowalski RA (1976) The semantics of predicate logic as a programming language. Journal of the ACM (JACM) 23(4):733–742

Propositional logic with probability operators (based on general ideas of weak modal calculus)

Tomasz Witczak

Abstract Our aim is to establish propositional calculus based on classical or intuitionistic core and equipped with probability operators. It is inspired also by the idea of weak modal logics. By that we mean systems which are not normal, regular or even monotonic. Roughly speaking, we consider pre-ordered and measurable relational structures of possible worlds where each world has its family of neighborhoods. The probability of a formula φ in a world w depends on measurability of $V(\varphi)$ and on whether $V(\varphi)$ belongs to $\mathcal{N}_w$. Thus, possible worlds different than the actual world play the role of "advisors", whose opinion can be accepted or not (this depends not only on the "strength" of this opinion but also on other, maybe very arbitrary reasons, which are described by neighborhoods). This work should be considered mostly as a draft of an overall research project although it contains not only basic ideas but also some results.

Key words: modal logic, other non-classical logics, intuitionistic logic, foundations of probability theory

1 Introduction

As for the intuitionistic modal logic, it has quite a long history (see Simpson (1994) for detailed survey). There exist various systems of that kind but usually they are rather strong (e.g. *normal* in modal sense). Thus, some authors started to investigate weak modal logics based on intuitionistic core. Some initial results have been collected in

Tomasz Witczak
Institute of Mathematics
University of Silesia
Bankowa 14, 40-007 Katowice, Poland
e-mail: tm.witczak@gmail.com

Witczak (2018). Different but similar ideas can be found in Dalmonte et al (2018). Classical weak modal systems have been extensively studied in Indrzejczak (2007).

Neighborhood semantics is a typical tool for reasoning about weak modalities. It is because non-normal systems are not complete with respect to the standard Kripke structures but it is possible to establish neighborhood completeness for some of them. In fact, we are able to do it even for the system **E** which contains only instances of classical tautologies, rule *modus ponens* and rule of extensionality.

If we speak about probability logics, then our main inspiration came from various works published by a team of researchers from Serbia. Their results have been gathered in Ognjanović et al (2009) and Ognjanović et al (2016) but we refer also to Markovic et al (2003). It should be noted that Serbian logicians did not use neighborhoods but only relational structures with measure.

2 Classical modal cases

2.1 Alphabet and language

Following Markovic et al (2003), we make clear distinction between purely classical formulas (the set For_C) and formulas with probabilistic operator $P_{\geq s}$. The latter set is named For_P. Basically, we do not allow for nesting of operators and for mixing of formulas taken from these two sets.

Here we list basic components of our initial language **cl1** :

1. PV is a fixed denumerable set of propositional variables $p, q, r, s, \ldots$
2. Logical connectives and operators are $\wedge, \vee, \rightarrow, \bot, \neg$ and $P_{\geq s}$, where s belongs to $S = \mathbb{Q} \cap [0, 1]$.

Small Greek letters $\varphi, \psi, \gamma, \alpha, \beta \ldots$ denote classical formulas (without probabilistic operators) and big Latin letters $A, B, C, D, \ldots$ denote formulas from For_P. The set of *all* formulas (both classical and probabilistic) will be named *For*. Its elements are denoted by $\Phi, \Psi, \Upsilon \ldots$

Formulas are generated inductively. Each propositional variable belongs to For_C. If $\varphi, \psi \in For_C$, then also $\varphi \wedge \psi, \varphi \vee \psi, \varphi \rightarrow \psi$ and $\neg \varphi$ are in For_C. If $\varphi \in For_C$ and $s \in S$, then $P_{\geq s}\varphi$ is named *basic probability formula (bpf)*. The set of all probabilistic formulas is named For_P. It is the smallest set that contains all *bpf*'s and is closed under the formation rules: if $A, B \in For_P$, then $A \wedge B, A \vee B, A \rightarrow B, \neg A \in For_P$.

We can say that the alphabet of **cl1** is just the alphabet of **cl0** (i.e. classical propositional language) but with an additional operator P.

2.2 Classical probabilistic frames

Our initial structure is a neighborhood frame for classical modal logic (**cl1**-frame) defined as it follows:

Definition 2.1. **cl1**-*frame is a triple* $F = \langle W, \mathcal{N}, \mathcal{H}, \mu \rangle$ *where* $\mathcal{N}$ *is a function from W into* $P(P(W))$, $\mathcal{H}$ *is an algebra od subsets of W and* μ *is a finitely additive measure, i.e.* $\mu : \mathcal{H} \to [0, 1]$.

Sometimes we shall use shortcut "$w \in M$" to say that $M = \langle W, \mathcal{N}, \mathcal{H}, \mu \rangle$ and $w \in W$.

Having frame, we can introduce model:

Definition 2.2. **cl1**-*model is a quintuple* $M = \langle W, \mathcal{N}, H, \mu, V \rangle$, *where V is a function from PV into* $P(W)$.

Of course, we can also define V in the following way: $V : W \times PV \to [0, 1]$. As for the forcing of formulas in particular worlds, it is defined inductively:

Definition 2.3. *In each* **cl1**-*model forcing relation* $\Vdash$ *is defined as below:*

1. When $q, \varphi, \psi \in For_C$:
   - $w \Vdash q \Leftrightarrow w \in V(q)$
   - $w \Vdash \varphi \wedge \psi \Leftrightarrow w \Vdash \varphi$ and $w \Vdash \psi$
   - $w \Vdash \varphi \vee \psi \Leftrightarrow w \Vdash \varphi$ or $w \Vdash \psi$
   - $w \Vdash \varphi \to \psi \Leftrightarrow w \nVdash \varphi$ or $w \Vdash \psi$
   - $w \Vdash \neg \varphi \Leftrightarrow w \nVdash \varphi$

2. When $\varphi \in For_C$ and $A, B \in For_P$:
   - $w \Vdash P_{\geq s}\varphi \Leftrightarrow \mu(V(\varphi)) \geq s$ and $V(\varphi) \in \mathcal{N}_w, s \in S$.
   - The other cases are just like in For_C, e.g. $w \Vdash A \wedge B \Leftrightarrow w \Vdash A$ and $w \Vdash B$.

As usual, we say that a formula φ is *satisfied* in **cl1**-model M when it is forced in each world of this model. It is *true* when it is *satisfied* in each **cl1**-model.

2.3 Alternative notions of probability

Our first definition of probability was based on the standard definition of necessity in neighborhood semantics. More precisely, it combines ideas from Ognjanović et al (2016) with classical modal approach which is shown below:

$w \Vdash \Box \varphi \Leftrightarrow V(\varphi) \in \mathcal{N}_w$

In this section we shall show two different operators: P' and P''. They are similar to P but in some sense more complex. Moreover, we shall speak about $_TP$, $_TP'$ and $_TP''$ (special versions of the operators in question). Later we shall present non-mathematical interpretation of all those notions of probability.

Let us introduce some formal definitions in order to maintain precision.

Definition 2.4. *If our initial language is called* **cl0**, *then we define languages* **cl2**, **cl3**, **cl1T**, **cl2T** *and* **cl3T** *as* **cl0** *extended with the operator (resp.)* P', P'', $_TP'$ *and* $_TP''$.

Definition 2.5. *We define* **cl2** - **cl3** -, **cl1T** -, **cl2T** - *and* **cl3T** -*frame (model) as* **cl1** -*frame (model) but with forcing of* P' *(resp.* P'', $_TP$, $_TP'$ *and* $_TP''$*) stated as below (we assume that* $\varphi \in For_C$*):*

- $w \Vdash P'_{\geq s}\varphi \Leftrightarrow$ *there exists measurable* $X \in \mathcal{N}_w$ *such that* $\mu(X) \geq s$ *and for each* $z \in X$ *we have* $z \Vdash \varphi$
- $w \Vdash P''_{\geq s}\varphi \Leftrightarrow \mu(Y) \geq s$, *where measurable* $Y = \bigcup \mathcal{X} \in \mathcal{N}_w$ *and* $\mathcal{X} = \{X \in \mathcal{N}_w;$ *for each* $z \in X, z \Vdash \varphi\}$
- $w \Vdash {}_TP_{\geq s}\varphi \Leftrightarrow w \Vdash \varphi, \mu(V(\varphi)) \geq s$ *and* $V(\varphi) \in \mathcal{N}_w, s \in S$.
- $w \Vdash {}_TP'_{\geq s} \Leftrightarrow w \Vdash \varphi$ *and there exists measurable* $X \in \mathcal{N}_w$ *such that* $\mu(X) \geq s$ *and for each* $z \in X$ *we have* $z \Vdash \varphi$
- $w \Vdash {}_TP''_{\geq s}\varphi \Leftrightarrow w \Vdash \varphi$ *and* $\mu(Y) \geq s$, *where measurable* $Y = \bigcup \mathcal{X} \in \mathcal{N}_w$ *and* $\mathcal{X} = \{X \in \mathcal{N}_w;$ *for each* $z \in X, z \Vdash \varphi\}$

As we can see, in the last three cases we require forcing of formula φ already in w. Later we shall show that such restriction allows us to accept probabilistic version of the well-known modal axiom T.

2.4 Non-mathematical interpretation of operators

Each operator has its own non-mathematical interpretation and describes certain specific situation:

1. P: probability of formula φ is greater (or equal to) than s if the set of worlds satisfying φ has sufficient measure and also belongs to the space of w-neighborhoods.
2. P': probability of formula φ is greater (or equal to) than s if we can find sufficiently "big" (in terms of measure) set of worlds forcing φ and this set is somewhere among our neighborhoods, i.e. it is accepted from the "neighborhood point of view".
3. P'': we consider all neighborhoods satisfying φ. Then we take their union and check two things: if its measure is sufficient and whether this union is neighborhood. In some sense, we speak about maximal neighborhood satisfying φ.
4. $_TP$, $_TP'$ and $_TP''$: we say that w must force φ and then we check if φ is "accepted" by means of measure and neighborhoods. Thus, probability becomes something like an additional confirmation of formula. Possible worlds are like our advisory board.

2.5 Towards axiomatization and completeness

In this section we limit ourselves to the basic operator P, i.e. we work in language **cl1** and with **cl1** -models. First, we shall check certain axioms which can be considered

Propositional logic with probability operators

as intuitively true formulas describing well-known properties of probability. In fact, some of them are not satisfied in our environment. We shall not show full proofs and countermodels, rather their sketches only (and some suggestions).

- A1. $P_{\geq 0}\alpha$. This very natural axiom is *not true* in **cl1** -model. Shortly speaking, it is very easy to imagine model M with a world w such that $V(\varphi) \notin \mathcal{N}_w$. This is enough to say that $w \not\Vdash P_{\geq 0}\alpha$.
- A2. $P_{\leq r}\alpha \to P_{<s}\alpha, s > r$. This axiom is obviously true (because of the basic features of measure).
- A3. $P_{<s}\alpha \to P_{\leq s}\alpha$. This axiom is also satisfied.
- A4. $(P_{\geq r}\alpha \land P_{\geq s}\beta \land P_{\geq 1}(\neg(\alpha \land \beta))) \to P_{\geq \min(1, r+s)}\alpha \lor \beta$.

  Suppose that there are M and $w \in M$ such that w satisfies the left side of the formula in question. Then $\mu(V(\alpha)) \geq r, \mu(V(\beta)) \geq s, V(\alpha) \in \mathcal{N}_w, V(\beta) \in \mathcal{N}_w, V(\neg(\alpha \land \beta)) \geq 1$ and $V(\neg(\alpha \land \beta)) \notin \mathcal{N}_w$. Now assume that the right side is not true in w. Hence, $\mu(V(\alpha \lor \beta)) < \min(1, r+s)$ or $V(\alpha \lor \beta) \notin \mathcal{N}_w$. The first part of this disjunction is clearly impossible by the very properties of measure. However, the second is possible.

  Note. There is an easy way to accept this axiom. We should assume that our neighborhoods are closed under (at least finite) unions: $X, Y \in \mathcal{N}_w \Rightarrow X \cup Y \in \mathcal{N}_w$ (*). Then $V(\alpha \lor \beta) = V(\alpha) \cup V(\beta) \in \mathcal{N}_w$. The axiom in question corresponds to the additivity of measures (see Ognjanović et al (2009) and Ognjanović et al (2016)). In fact, it means that the lower bound of $\mu(V(\alpha \lor \beta))$ cannot be lesser than $\mu(V(\alpha)) + \mu(V(\beta))$ (for *disjoint* $V(\alpha), V(\beta)$).

- A5. $(P_{\leq r}\alpha \land P_{<s}\beta) \to P_{<r+s}(\alpha \lor \beta), r + s \leq 1$.

  Suppose that there are $M, w \in W$ such that w satisfies the left side of the formula in question. Then $\mu(V(\alpha)) \leq r, \mu(V(\beta)) < s, V(\alpha) \in \mathcal{N}_w$ and $V(\beta) \in \mathcal{N}_w$. Now assume that the right side is not true in w. Hence, $\mu(V(\alpha \lor \beta)) \geq r + s$ or $V(\alpha \lor \beta) \notin \mathcal{N}_w$. As earlier, the first part of this disjunction is impossible but the second is possible.

  Note. Again, we can easily accept this axiom, assuming that (*) holds. The axiom says that the upper bound of $\mu(V(\alpha \lor \beta))$ cannot be greater than $\mu(V(\alpha)) + \mu(V\beta))$ (for disjoint $V(\alpha), V(\beta)$).

- A6. $\Phi, \Phi \to \Psi \vdash \Psi$. This is well-known rule *modus ponens* and it is accepted.
- A7. $\alpha \vdash P_{\geq 1}\alpha$. This rule (in some sense similar to the modal rule of necessity) is not accepted. We can easily imagine that α is a tautology (thus it is accepted in each world of each model) but at the same time there is a model M with a world w such that $V(\alpha) \notin \mathcal{N}_w$. Hence, we cannot say that $P_{\geq 1}\varphi$ is true.
- A8. $A \to P_{\geq s - 1/k}\alpha, k \geq 1/s, s > 0 \vdash A \to P_{\geq s}\alpha$

  Suppose that for each world v in each model there is $v \not\Vdash A$ or $v \Vdash P_{\geq s-1/k}\alpha$. Assume now that there are $M, w \in M$ such that $w \not\Vdash A \to P_{\geq s}\alpha$. Hence, $w \Vdash A$ and $w \not\Vdash P_{\geq s}\alpha$. Thus $\mu(V(\alpha)) < s$ or $V(\alpha) \notin \mathcal{N}_w$. But we know that for each $k \geq 1/s, s \in S$ we have $(\mu(V(\alpha)) \geq s - 1/k$ and $V(\alpha) \in \mathcal{N}_w$. Finally, the rule in question holds.

- A9. $P_{\geq 1}(\alpha \to \beta) \to (P_{\geq s}\alpha \to P_{\geq s}\beta)$

  Note. Let us agree on the additional frame condition: $X \in \mathcal{N}_w \Rightarrow -X \notin \mathcal{N}_w$ (**)

Suppose that there are $M, w \in M$ such that $w \Vdash P_{\geq 1}(\alpha \to \beta)$ and $w \nVdash P_{\geq s}\alpha \to P_{\geq s}\beta$. Let us decompose and rewrite all those conditions: c_1) $\mu(V(\neg\alpha) \cup V(\beta)) \geq 1$
and [c_2) $V(\neg\alpha) \in \mathcal{N}_w$ or c_3) $V(\beta) \in \mathcal{N}_w$]
and c_3) $\mu(V(\alpha)) \geq s$
and c_4) $V(\alpha) \in \mathcal{N}_w$
and [c_5) $\mu(V(\beta)) < s$ or c_6) $V(\beta) \notin \mathcal{N}_w$].

Thus we can be sure that c_1, c_3 and c_4 hold. Suppose that we accept c_2. Now we use (**) and the fact that $V(\neg\alpha) = -V(\alpha)$ to say that $V(\alpha) \notin \mathcal{N}_w$. Contradiction with c_4. Now let us accept c_6 and c_3. It is clear contradiction. Thus we must check conjunction of c_5 and c_3. Take $s = \frac{1}{3}$. Of course, by the very properties of probability, $\mu(V(\neg\alpha) \cup V(\beta)) \leq \mu(V(\neg\alpha)) + \mu(V(\beta))$. If $\mu(V(\alpha)) \geq s$, then $\mu(-V(\alpha)) = \mu(V(\neg\alpha)) < s$. Now $\mu(V(\neg\alpha)) + \mu(V(\beta)) < s + s = 2s = \frac{2}{3}$. This is contradiction with c_1.

Finally, we can accept the axiom in question: at least with an additional condition (**). Note that A9 is similar to the well-known modal axiom $K: \Box(\varphi \to \psi) \to (\Box\varphi \to \Box\psi)$. In Ognjanović et al (2009) and Ognjanović et al (2016) A9 was derived from other axioms but by means of A7.

A10. $\alpha \leftrightarrow \beta \vdash P_{\geq s}\alpha \leftrightarrow P_{\geq s}\beta$
This can be considered as the *rule of extensionality* (for the probability operator). It is easy to check that this rule is true.

A11. $P_{\geq 1}(\alpha \leftrightarrow \beta) \to (P_{=s}\alpha \to P_{=s}\beta)$.
Surprisingly, this axiom is not true. Let us imagine the following situation: we have model M, world $w \in M$ and formulas α, β such that $V(\alpha) \neq V(\beta), \mu(V(\alpha)) = \mu(V(\beta)), V(\alpha) \in \mathcal{N}_w$ and $V(\beta) \notin \mathcal{N}_w$ (thus $w \nVdash P_{=s}\beta$). Then it is still possible that $\mu(V(\alpha \leftrightarrow \beta)) \geq 1$ (even if $\mu(W) = 1$). It is because $V(\alpha)$ and $V(\beta)$ can differ on the set of measure zero.

Now let us consider system *PROB* which is based on the language **cl1** and contains: all instances of the classical propositional tautologies, A2, A3, A6, A8, A9 and A10. We say, following Ognjanović et al (2009), that formula Ψ is *deducible from a set* T of formulas ($T \vdash \Psi$) if there exists an at most denumerable sequence of formulas $\Psi_0, \Psi_1, ..., \Psi$, such that every Ψ_i is an axiom or a formula from T, or it is derived from the preceding formulas by means of an inference rule. Such sequence is called a *proof* of Ψ from T. *Theorems* are formulas deducible from empty set (more precisely: from the set of axioms and rules).

The next theorem should be considered as a statement about *PROB*.

Theorem 2.6. *If T is a set of formulas and $\varphi, \psi \in For_C$ or $\varphi, \psi \in For_P$, then:* $T \cup \{\varphi\} \vdash \varphi \to \psi$.

Proof. (sketch)
Basically, we can repeat the proof of Theorem 9 from Ognjanović et al (2009). It means that implication from right to left can be realized just as in classical propositional case. For the other direction we use transfinite direction on the length of proof of ψ from $T \cup \{\varphi\}$. There is one difference: we do not have "rule 2" (A7 in our notation). But

Propositional logic with probability operators

this is not a problem: the only thing to do is to reject this case. As for the case of "rule 3" (A8) it does not depend on A7.

□

A short discussion of the completeness issue. In fact, at this moment we are not able to prove completeness of *PROB* or any other system based on our earlier considerations. We have some clues but also some problems which should be solved.

First of all, the definition of *maximal consistent* set T in Ognjanović et al (2009) requires the following conditions: 1) for every $\alpha \in For_C$, if $T \vdash \alpha$, then $\alpha \in T$ and $P_{\geq 1}\alpha \in T$; 2) for every $A \in For_P$, either $A \in T$ or $\neg A \in T$. It seems that we can adapt this definition to prove some important lemmas:

Lemma 2.7. *Assume that T is a consistent set of formulas. Then:*
 1) for any formula $A \in For_P$, either $T \cup \{A\}$ is consistent or $T \cup \{\neg A\}$ is consistent.
 2) If $\neg (\alpha \to P_{\geq s}\beta) \in T$, then there is some $n > \frac{1}{s}$ such that $T \cup \{\alpha \to \neg P_{\geq s-\frac{1}{n}}\beta\}$ is consistent.

Lemma 2.8. *Suppose that T is a maximal consistent set of formulas. Then:*

1. *for each formula $A \in For_P$, exactly one member of $\{A, \neg A\}$ belongs to T.*
2. *for all formulas $A, B \in For_P$, $A \vee B \in T \Leftrightarrow A \in T$ or $B \in T$.*
3. *for all formulas φ, ψ, where either $\varphi, \psi \in For_C$ or $\varphi, \psi \in For_P$, $\varphi \wedge \psi \in T \Leftrightarrow \{\varphi, \psi\} \subseteq T$.*
4. *for each $\Psi \in For$, if $T \vdash \varphi$, then $\varphi \in T$.*
5. *for all formulas φ, ψ, where either $\varphi, \psi \in For_C$ or $\varphi, \psi \in For_P$, if $\{\varphi, \varphi \to \psi\} \subseteq T$, then $\psi \in T$.*
6. *for all formulas φ, ψ, where either $\varphi, \psi \in For_C$ or $\varphi, \psi \in For_P$, if $\varphi \in T$ and $\vdash \varphi \to \psi$, then $\psi \in T$.*
7. *for any formula α, if $t = \sup_s \{P_{\geq s}\alpha \in T\}$ and $t \in S$, then $P_{\geq t}\alpha \in T$.*

Note. It seems that first six properties can be proved just like in a typical classical propositional case. For the last property we need only monotonicity of the measure (but we have it) and rule A8.

Lemma 2.9. *(Lindenbaum lemma). Every consistent set can be extended to a maximal consistent set.*

Note. Here we are suggested by Ognjanović et al (2009) and Ognjanović et al (2016) to perform certain procedure quite similar to the standard one. In other words, we should define a sequence of sets T_i, satisfying various properties and then we should show that $\mathcal{T} = \bigcup_{i=0}^{\infty} T_i$. The proof does not require any specific axiom or rule which would be absent in our initial *PROB*-package.

As for the canonical model, it is based (again see Ognjanović et al (2009)) on the $W = \{w \models Cn_C(T)\}$, where each w is a classical propositional interpreation satisfying the set $Cn_C(T)$ of all classical consequences of the consistent T. Then we define $[\alpha]$ as $\{w \in W; \ w \models \alpha\}$ and H as $\{[\alpha] : \alpha \in For_C\}$. As for the measure, it is a function $\mu : H \to [0, 1]$ such that $\mu([\alpha]) = \sup_s \{P_{\geq s}\alpha \in \mathcal{T}\}$. Here we have the first problem: without neighborhoods, such definition allows us to say that $\mu([\alpha]) \geq 0$. It is because of axiom A1 which is accepted in the original system of Ognjanović, Rašković and Marković. In our environment it is not accepted and we can even say that it *should not* be accepted (if we want to treat neighborhoods as an additional, arbitrary criterion which allows us to accept or deny probabilistic formula). One solution which could be checked, is to assume that:

$$\mu([\alpha]) = \begin{cases} \sup_s \{P_{\geq}\alpha \in \mathcal{T}\}; & \text{if there is any } z \in S \text{ for which } P_{\geq z}\alpha \in \mathcal{T} \\ 0; & \text{else} \end{cases}$$

Moreover, it is natural to expect that $\mu([\alpha] \cup [\beta]) = \mu([\alpha]) + \mu([\beta])$ (for all disjoint $[\alpha]$ and $[\beta]$). For this reason we should accept A4, A5 and - on the semantical side - condition (*). Also, we need condition (**) to obtain correspondence with A9. Now the problem is: how to define neighborhoods in canonical model in such a way that it would satisfy (*) and (**)? Basically, we think that it would be sensible to start from the following definition: $X \in \mathcal{N}_w \Leftrightarrow \exists \varphi \in For_C, s \in S$ such that: if $P_{\geq s}\varphi \in w$, then $\varphi \in v$ for each theory $v \in X$. This will be a matter of further research.

3 Intuitionistic cases

In this section we present short introduction to the probabilistic relational-neighborhood frames with monotonicity of forcing. Thus, our system becomes intuitionistic.

3.1 Intuitionistic probabilistic frames

Following Markovic et al (2003), we make clear distinction between purely intuitionistic formulas (the set For_I, which, in particular, contains propositional variables) and formulas with probabilistic operators $P_{\leq s}, P_{\geq s}$. The latter set is named $_IFor_P$. Again, we do not allow for nesting of operators and for mixing of formulas taken from these two sets.

We shall present three classes of structures and four approaches to the forcing of probabilistic formulas. We start from a very simple frame which is not useful in itself but can be considered as a foundation for more complex objects.

This initial structure is a pre-ordered neighborhood frame for intuitionistic modal logic (**pn**-frame) defined as it follows:

Definition 3.1. pn *-frame is a triple* $\langle W, \mathcal{N}, \leq \rangle$ *where* $\leq$ *is a partial order on W and* $\mathcal{N}$ *is a function from W into* $P(P(W))$.

As we announced, this definition is not very helpful because it does not give us any connection between neighborhoods and pre-order. Thus, we introduce two subclasses:

Definition 3.2. inp1 *-frame is a* **pn** *-frame where:*

$$w \leq v \Rightarrow \mathcal{N}_w \subseteq \mathcal{N}_v \qquad (1)$$

Definition 3.3. inp2 *-frame is a **pn** -frame where:*

$$w \leq v, v \in X \subseteq W, X \in \mathcal{N}_w \Rightarrow X \in \mathcal{N}_v \qquad (2)$$

Clearly, condition (1), typical for **inp1** -frames, implies condition (2) (but they are not equivalent).

3.2 From frames to models

Having frames, we can introduce models.

Definition 3.4. inp1 *(resp.* **inp2** *)-model is a sextuple* $\langle W, \mathcal{N}, \leq, H, \mu, V \rangle$*, where:*

1. $\langle W, \mathcal{N}, \leq \rangle$ is an **inp1** *(resp.* **inp2** *)-frame,*
2. $\langle W, \leq, V \rangle$ is an intuitionistic model (here we restrict forcing of formulas to For_I),
3. H is the smallest algebra on W (containing W, closed under complementation and finite union), which contains $H_I = \{V(\varphi);\ \varphi \in For_I\}$ and $H'_I = \{W \setminus V(\varphi);\ \varphi \in For_I\}$ (see Markovic et al (2003)).
4. $\mu : H \rightarrow S$ is a finitely additive probability, S is a recursive subset of $[0,1]$ containing all rational numbers from this interval (see Markovic et al (2003)).

As for the forcing of formulas, it is defined inductively. We said earlier that we shall present four approaches to the forcing of probabilistic formulas. Thus, for convenience, we introduce four probabilistic operators: P, $_TP$, P' and P''.

Definition 3.5. *In each* **inp1** *- and* **inp2** *-model forcing of formulas without probabilistic operators is just like in any standard intuitionistic model. In particular:*
$w \Vdash \varphi \rightarrow \psi \Leftrightarrow$ *for each* $v \in W$*, such that* $w \leq v, v \nVdash \varphi$ *or* $v \Vdash \psi$*.*
Forcing of probabilistic formulas is defined in the following way:

- $w \Vdash P_{\geq s}\varphi \Leftrightarrow \mu(V(\varphi)) \geq s$ and $V(\varphi) \in \mathcal{N}_w, s \in S$ (in **inp1** -models).
- $w \Vdash {_TP_{\geq s}}\varphi \Leftrightarrow w \Vdash \varphi, \mu(V(\varphi)) \geq s$ and $V(\varphi) \in \mathcal{N}_w$ (in **inp1** - and **inp2** -models)
- $w \Vdash P'_{\geq s}\varphi \Leftrightarrow$ there exists measurable $X \in \mathcal{N}_w$ such that $\mu(X) \geq s$ and for each $z \in X$ we have $z \Vdash \varphi$ (in **inp1** -models).
- $w \Vdash P''_{\geq s}\varphi \Leftrightarrow \mu(Y) \geq s$, where measurable $Y = \bigcup \mathcal{X}$, where $\mathcal{X} = \{X \in \mathcal{N}_w;\ \text{for each } z \in X, z \Vdash \varphi\}$ and $Y \in \mathcal{N}_w$ (in **inp1** -models).

One could say that we can modify definitions of P' and P'', making them more similar to the definition of $_TP$. For this we should add simple condition: that $w \Vdash \varphi$. Of course such modification is possible but we shall not dwell on this topic here. Another requirement would be to expect that:
$w \Vdash P'''_{\geq s}\varphi, s \in S \Leftrightarrow \mu(V(\varphi)) \geq s$ and $V(\varphi) \in \mathcal{N}_v$ for each $v \in W$ such that $w \geq v$.

Non-mathematical interpretation of those operators is similar for the classical case. In general, it means that worlds from w-neighborhoods are "advisors" for w. The difference is that here we have also pre-order and monotonicity of forcing. This feature shall be proven later. Now let us discuss another problem.

3.3 Negation

There is an interesting question of negation. For short, let us denote formulas from $_IFor_P$ by capitals: A, B, C.... We can define negation of A in a manner similar to Markovic et al (2003), i.e. in classical way:

$w \Vdash \sim A \Leftrightarrow w \nVdash A$

On the other hand, we can adopt intuitionistic notion:

$w \Vdash \neg A \Leftrightarrow$ for each v such that $w \leq v$, $v \nVdash A$

One can easily show that $A \vee \sim A$ is tautology but $A \vee \neg A$ not. Also implication between $_IFor_P$-formulas can be classical ($\rightsquigarrow$) or intuitionistic ($\rightarrow$). Basically, we shall work with $\neg$ and $\rightarrow$.

3.4 Monotonicity of forcing

If we speak about intuitionistic models, then it is obvious that forcing of formulas should be monotone with respect to the pre-order. Let us prove the following lemmas:

Lemma 3.6. *In each* **inp1** *-model* $M = \langle W, \mathcal{N}, \leq, H, \mu, V \rangle$: *if* $w \Vdash \varphi$ *and* $w \leq v$, *then* $v \Vdash \varphi$.

Proof. The only problematic case is that of probabilistic formulas.

Assume that $\varphi = P_{\geq s}\alpha$, where $\alpha \in For_I$. As we said, $w \Vdash P_{\geq s}\alpha$. Thus $\mu(V(\varphi)) \geq s$ and $V(\varphi) \in \mathcal{N}_w$. If $w \leq v$, then $V(\varphi) \in \mathcal{N}_v$. Thus $v \Vdash \varphi$.

Now assume that $\varphi = P'_{\geq s}\alpha$, where $\alpha \in For_I$. Suppose that $w \Vdash P'_{\geq s}$. Thus there is measurable $X \in \mathcal{N}_w$ such that $\mu(X) \geq s$ and for each $z \in X$ we have $z \Vdash \varphi$. If $w \leq v$, then $X \in \mathcal{N}_v$ and thus $v \Vdash \varphi$.

Proof for P'' is very similar. □

Lemma 3.7. *In each* **inp2** *-model* $M = \langle W, \mathcal{N}, \leq, H, \mu, V \rangle$: *if* $w \Vdash \varphi$ *and* $w \leq v$, *then* $v \Vdash \varphi$.

Proof. Here we work only with $_TP$ operator. Assume that $\varphi = {}_TP_{\geq s}\alpha$. Hence, $w \Vdash \varphi$, $\mu(V(\varphi)) \geq s$ and $V(\varphi) \in \mathcal{N}_w$. Thus, by induction hypothesis, if $w \leq v$, then $v \Vdash \varphi$. Now $v \in X = V(\alpha) \in \mathcal{N}_w$. Hence, $V(\alpha) \in \mathcal{N}_v$. Finally, we get $v \Vdash \varphi$. □

4 Final remarks

As for the axiomatization of intuitionistic system, we can say that almost all considerations about probabilistic axioms are identical with the classical ones. Of course we must remember that "inside" our logic we do not have classical calculus but the intuitionistic system without the law of excluded middle.

In general, it would be fruitful to obtain completeness for the classical system *PROB* (with axioms A4 and A5) with respect to **cl1**-frames (with additional requirements (*) and (**)). Moreover, there is a question (solved in Ognjanović et al (2009) but clearly without neighborhoods) of nesting probabilistic operators and mixing classical formulas with those from For_P.

Our goal was to show that the usage of neighborhoods gives us certain benefits: we can speak about *restricted* probability. We have two levels of verification: the first depends on the measure (of $V(\alpha)$) and the second on the belonging of $V(\alpha)$ to the family of neighborhoods.

References

Dalmonte T, Grellois C, Olivetti N (2018) Toward intuitionistic non-normal modal logic and its calculi

Indrzejczak A (2007) Labelled tableau calculi for weak modal logics. Bulletin of the Section of logic 36(3-4):159–173

Markovic Z, Ognjanovic Z, Rašković M (2003) An intuitionistic logic with probabilistic operators. Publications de L'Institute Matematique, ns 73(87):31–38

Ognjanović Z, Rašković M, Marković Z (2009) Probability logics. In: Logic in computer science, Springer, pp 35–111

Ognjanović Z, Rašković M, Marković Z (2016) Probability logics: probability-based formalization of uncertain reasoning. Springer

Simpson AK (1994) The proof theory and semantics of intuitionistic modal logic

Witczak T (2018) Simple example of weak modal logic based on intuitionistic core. arXiv preprint arXiv:180609443

Let Me Ask You an Easier Question—Modifying and Rephrasing Questions in Information Seeking Dialogues*

Paweł Łupkowski

Abstract The main aim of this paper is to explore and analyse how questions are modified and rephrased in information seeking dialogues. I am interested in situations where a question is modified by one of a dialogue participants in order to facilitate the answering process. Examples of such questions' modifications are retrieved from natural language corpora: Erotetic Reasoning Corpus (a data set for research on natural question processing), The Basic Electricity and Electronics Corpus (tutorial dialogues from electronics courses) and The British National Corpus.

Key words: questions, query-responses, question dependency, logic of questions, decomposition principle, topicality

Introduction

In a dialogue, especially in an information seeking dialogue, we encounter situations when a query is not followed by a simple answer, but by another query. As discussed in (Łupkowski and Ginzburg, 2013, 2016) and (Ginzburg et al, 2019) such a query-response may take different forms and come with different dialogue participant's motivations. Query-responses typology covers the following types: (1) CR: clarification requests; (2) DP: dependent questions, i.e. cases where the answer to the initial question depends on the answer to a q-response; (3) MOTIV: questions about an underlying motivation behind asking the initial question; (4) NO ANSW: questions aimed at

---

Paweł Łupkowski
Department of Logic and Cognitive Science
Faculty of Psychology and Cognitive Science
Adam Mickiewicz University, Poznań, Poland
e-mail: `Pawel.Lupkowski@amu.edu.pl`

* This work was supported by funds of the National Science Centre, Poland (DEC-2012/04/A/HS1/-00715).

avoiding answering the initial question; (5) FORM: questions considering the way of answering the initial question; (6) QA: questions with a presupposed answer, (7) IGNORE: responses ignoring the initial question—for more details see (Łupkowski and Ginzburg, 2016, p. 355). One of the most interesting query-responses are these which facilitate answering to the initial questions. Such query-responses may be analysed in the light of the so called *erotetic decomposition principle* (Wiśniewski, 2013, p. 103).

EDP (*Erotetic Decomposition Principle*) Transform a principal question into auxiliary questions in such a way that: (a) consecutive auxiliary questions are dependent upon the previous questions and, possibly, answers to previous auxiliary questions, and (b) once auxiliary questions are resolved, the principal question is resolved as well.

Let us illustrate such a transformation with the following example retrieved from the British National Corpus (Burnard, 2007):

(1) A: Do you want me to <*pause*> push it round?
    B: Is it really disturbing you?
    [*BNC:FM1*, 679–680][2]

What is the reason behind B's decision to respond with a question to A's question in this case? Why B would not simply provide an answer to A's question? B has a piece of information that allows to process A's question and to produce a dependent auxiliary question (cf. *Whether I want you to push it depends on whether it really disturbs you*)[3]. This additional information is something that B probably accepts in this dialogue situation (but is not explicitly given by B). One of the possibilities in this case might be: "I want you to push it round if and only if it is disturbing you". If we accept this premise, our example will appear as below. We may consider this additional information to play the role of enthymematic premise in the analysed erotetic inference[4]:

(2) A: Do you want me to <*pause*> push it round?
    B: *I want you to push it round iff it is disturbing you.*
    B: Is it really disturbing you?

In the situation presented above A's question might be interpreted as an expression of a problem/issue to be solved. In other words, after the announcement of the question "Do you want me to <*pause*> push it round?" it becomes a problem to be solved by B. We may say, that this question becomes B's initial question that has to be processed. Such an interpretation allows us to grasp the rationale behind B's dialogue move in the example. Now we are focused only on B's side, because the question about pushing

---

[2] This notation indicates the sentences numbers (679–680) of a BNC file (FM1). Original spelling is preserved in the examples.

[3] $q1$ **depends on** $q2$ iff any proposition p such that p **resolves** $q2$, also satisfies p entails r such that r **is about** $q1$. (Ginzburg, 2012, (61b), p. 57); see also (Łupkowski, 2016, Chapter 3).

[4] See (Wiśniewski, 2013, p. 51–52). See also a broader discussion concerning modelling natural language examples in Inferential Erotetic Logic (IEL) in (Łupkowski, 2016).

the thing around becomes B's initial question. After the initial question is processed B comes to the question "Is it really disturbing you?", which is later asked to A.

What is characteristic for the discussed example is that the initial question is asked by one of the dialogue participants (A) and a question-response is provided by the other one (B).[5] In this paper I will focus on situations where *the same* dialogue participant replaces the initial question with another one. Rationale underlying mechanisms behind such a dialogue step is usually analogical to the one described above, i.e. to facilitate the answering process for the initial question. What is more, such cases are more intuitive for analysis in the light of EDP as they shed light on the process of questions' transformations done by the questioning agent. The motivational example for this study is presented below.

(3) A: Question six (pause) okay for anybody who's interested in eating, as we are, *pate de foie gras* is made from what?
 A: Right we'll be even more specific right, a help for ya, *pate de foie gras* is made from the liver of what?
 [*BNC: KDC*, 20–21]

As it is visible in the example, A asks the initial question and in what follows s/he replaces it with another one. What is interesting, s/he clearly states the intention behind this move: "a help for ya". What is even more, s/he points out why her/his second question should facilitate the answering process: "we'll be even more specific right". The first question is more general than the second one, and one would agree that it is easier to answer the second question.

For the analysis presented in this paper one more notion will be useful, namely the *topic* in the meaning introduced by Van Kuppevelt (1995). *Topicality* is the general organising principle in discourse. The topic (for a discourse unit) is provided by an explicit or implicit question. Van Kuppevelt does not consider simple question/query-response pairs, but rather speaks about discourse units. However, the relation between such units is determined by the relation between the previously mentioned topic-providing questions. Thus we may say that the initial question from the motivational example sets a topic (i.e. the main ingredient of *pate de foie gras*). As Van Kuppevelt (1995, p. 111) puts it: "topic (...) refer[s] to a topic notion which concerns the 'aboutness' of (sets of) utterances". The modified question stays within the established topic, however it is (usually) more detailed.[6]

One more comment is needed here. In most cases (also the ones analysed here) the modification trigger is not directly observable in the corpus data. What we can do is to try to infer it from the broader context of the situation. We can imagine that in example (3) after asking the initial question A is confronted with silence, or maybe it is a face expression of her/his listeners that triggers the second question. It is a non-linguistic

---

[5] More such examples and formal models of question dependency may be found in (Łupkowski and Ginzburg, 2016) and (Łupkowski, 2016, 2017).

[6] It is worth to mention that the approach employing topicality and the situational semantics was successfully used for the modelling of linguistic data of the TZ sub-corpus of Erotetic Reasoning Corpus—see Urbański and Żyluk (2018), Urbański et al (2016) and *On two simple models for one simple game: 'Guess Who?', Inferential Erotetic Logic, and situational semantics* (this volume).

factor that allows A to understand that the initial question is to difficult or to complex to be answered. This suggests analysing the corpus data—as it is presented in this paper—is only a starting point for deeper research on the phenomenon of modifying and rephrasing questions. To grasp more understanding of the processes involved in triggering question modification additional linguistic and non-linguistic data will be needed. In my opinion however, the analysis of the corpus data offers a good and useful ramification for such future studies. I will also present some cases when the modification of the initial question is triggered by an utterance of other dialogue participant(s) and as such it is visible on the level of textual corpus data.

In my opinion, the importance of research on query-responses and the proposed research on modifying questions lays in the fact that their fit to a broader scheme of research concerning the issue of relevance of questions in the dialogical context—see wider discussion in (Ginzburg, 2012).

In what follows I would like to firstly, focus on the modification and rephrasing types observed in the data and secondly, analyse the issue of question modification triggers.

As the data source I use three language corpora containing different dialogue types.

1. The British National Corpus (BNC, Burnard 2007), large, balanced corpus containing free conversations. Allows for analysing general conversations and different contexts.
2. The Basic Electricity and Electronics Corpus (BEE; Rosé et al 1999), which contains tutorial dialogues from electronics courses. Goal directed conversations in the educational context.
3. The Erotetic Reasoning Corpus (ERC, Łupkowski et al 2018b), which constitutes a data set for research on natural question processing. The corpus consists of the language data collected in the previous studies on the question processing phenomenon. Goal directed conversations in the problem solving context. This source is especially valuable, because certain rephrasing situations are directly motivated by the second dialogue participant (and it is visible on the level of corpus data).

For the purpose of this paper the data from BNC and BEE corpora were retrieved with the use of the *SCoRE* tool (Purver, 2001), and the data from ERC were retrieved with the use of the *ERC Search & Browse Tool* (Marciniak and Łupkowski, 2017).

1 Question modification and rephrasing types

Let us now take a closer look at the gathered modification examples starting from the most general conversations retrieved from BNC, then moving for BEE and finally to ERC corpus. In order to simplify the description I will refer to the initial question as $q1$ and for the modified one as $q2$.

1.1 The British National Corpus

Open to open, but more specific question (O–O)

The motivational example (3) comes from a casual conversation, where dialogue participants play a form of a guessing game. Such a context seems to be natural for question modifications to appear. There are also other contexts where such a dialogue step is quite natural, like interviews. Let us consider such dialogue fragments.

(4) ANON 1: Were there any things about the job you didn't like?
    ANON 1: Any things you didn't like doing or felt uneasy doing?
    [*BNC: K65*, 943–944]

In the example above we observe that $q1$ and $q2$ are open questions, just like in the motivational example. It is also the case that $q2$ asks for more specific issues than $q1$. What is crucial is that we may say that $q2$ stays within the topic established by $q1$. The initial question considers a general "things" about a given job, which makes it rather difficult to answer straightforwardly. For $q2$ we observe that general "things" are replaced with more specific aspect of the job, i.e. "things you didn't like doing or felt uneasy doing". It is easier now for the respondent to identify these aspects of the job under discussion. The analysed example is a fragment of an interview. Such a dialogue move is important for the interviewing context, because the role of a person who is coordinating such a dialogue is to facilitate the answering process—it is important that the person who is asked questions during interview finds them easy to answer.

(5) ANON 1: The Chairman of the County Council has just got food poisoning (pause) Where did he pick it up?
    ANON 1: What restaurant was he in?
    ANON 1: Have you ever inspected the restaurant?
    [*BNC: KRP*, 250–252]

The example (5) is also a part of an interview. As before, $q1$ and $q2$ are open questions and it is the case where $q2$ asks for more specific issue within the range established by $q1$: "where" vs. "what restaurant".

Open to quasi-open (option suggested, O–QO)

In the following two examples (also being a part of an interview) we may observe modifying open question to open question, or rather quasi-open by introducing certain option in $q2$. It is not a full list of options/possibilities offered, but a suggestion to start from (in (6) "college" and in (7) "cleaning"). The respondent may choose the suggested option or any other which is suitable for s/he (which is encouraged by adding "or" at the end of $q2$).

(6) ANON 1: Where did you train?

ANON 1: Did you have to go to college or?
[*BNC*: *K*64, 9–10]

(7) ANON 1: Did he do anything else special with his knives?
ANON 1: <pause> When he was cleaning them or?
[*BNC*: *K*65, 602–603]

Open to choice (closed list of options, O–CH)

A somehow natural extension of the modifying strategy identified in (6) and (7) may be observed in examples (8), (9) and (10). Here $q1$ is an open question, while $q2$ simply offers list of proposed options to choose. Intuitively we may say that $q1$ introduces a topic, allows for focusing an a given issue and $q2$ calls for pointing a specific information about this issue—e.g. in (8) the initial question introduces issue: "what was life on the boat" and $q2$ calls for specific answer: was it "comfortable" or "bit basic".

(8) ANON 1: What was life like on the boat?
ANON 1: Was it quite comfortable or a bit basic?
[*BNC*: *K*64, 180–181]

(9) YVONNE: What would you have done with him if you had been able to arrest him?
YVONNE: As he was mental, I mean would you have taken him to a hospital or the station?
[*BNC*: *K*68, 345–346]

(10) JOHN: Okay, so what does sixteen to the power a half mean?
JOHN: What is or what value has it got?
[*BNC*: *K*6J, 306–307]

One may expect that also cases where the open $q1$ will be replaced with even more specific $q2$, like a yes/no question would also occur. We will observe them in a tutorial context in data retrieved from BEE corpus.

1.2 The Basic Electricity and Electronics Corpus

The BEE corpus consists of the goal-directed tutorial dialogues, thus questioning and answering play a crucial role.

Open to open, but more specific question (O–O)

What we observe are question modifications already described for BNC, namely a dialogue participant goes from open $q1$ to open (but more specific) $q2$. This is visible e.g. in (11) "what am I supposed to do", not generally but specifically with "one blue clip"; as well as in (12) "how to calculate voltage", not the process but the formula. Examples (13) and (14) are of the same character.

(11) STUDENT: what am I supposed to do?
STUDENT: how do I do it with one blue clip?
[BEE(F): Stud 28][7]
(12) TUTOR: And how do you calculate voltage?
TUTOR: what is the formula? [BEE(F): Stud 28]
(13) STUDENT: The things in the upper left hand corner.
TUTOR: Which things?
TUTOR: What do they look like?
STUDENT: They have colored lines in the middle and wires sticking out.
[BEE(F): Stud23]
(14) STUDENT: Because you need the physical object tand all of the connections inide of the body of the lamp to make it work
TUTOR: Why?
TUTOR: What would happen if you just had a lightbulb, wires, and an electricity source (like a battery)?
STUDENT: you would have a light but it wouldn't necessarally be a lamp.
[BEE(F): Stud25]

Choice (closed list of options to yes/no (CH–Y/N)

In the data we may also observe the case of modifying a closed choice $q1$ question into a yes-no $q2$, like in the example (15) below.

(15) TUTOR: And what would you say is creating the electricity?
STUDENT: the wires connected to the lightbulb
TUTOR: Are the wires creating the electricity or just allowing it to flow to the lightbulb?
TUTOR: If you had just a wire and a lightbulb, would you have enough to make it light?
STUDENT: no you also need a power source.
TUTOR: RIGHT!
TUTOR: You need a power source.
[BEE(F): Stud25]

---

[7] This notation indicates BEE sub-corpus (F—Final Experiment) and the file number (stud28). No sentence numbering is available for the BEE corpus. Original spelling is preserved in the examples.

Open to yes/no question (O–Y/N)

We also encounter example of question modification type suggested in the context of BNC data, i.e. $q1$ is an open question (why?) and $q2$ is a yes/no question.

(16) TUTOR: So would you predict that you could get a non-zero voltage reading across a switch or not?
STUDENT: I think that you could not get a non zero voltage reading.
TUTOR: Right, you are absolutely correct.
TUTOR: Can you explain why?
STUDENT: Because the switch itself is not directly giving the energy it is coming from the outlet.
TUTOR: OK, that is half the answer.
TUTOR: Think for a minute, why can you get a non-zero voltage reading across the lightbulb then?
TUTOR: Is it creating electricity?
[*BEE(F)*: *Stud*25]

In the example (16) above it is especially visible that $q2$ is asked here in order to facilitate the answering process for $q1$. When student will answer $q2$ it will help her/him to resolve the problem under $q1$. It is also the case that $q2$ is simpler to answer than $q1$.

1.3 The Erotetic Reasoning Corpus

For the analysis I use the TZ sub-corpus of ERC. ERC(TZ) consists of transcribed *Mind Maze* game sessions, thus we are dealing here with well structured conversation managed accordingly to a specific set of rules. *Mind Maze* is a card game published by Igrology. In the game one of the players is the game master and the other one tries to solve a puzzle presented by the game master. The game master tells a short story (inspired by true events) and the objective of the player is to figure out how the story happened by asking questions to the game master. Only yes/no questions are allowed here (with two additional admissible answers: "It is not important/relevant" and "It is not known")—(see Urbański et al, 2016; Urbański and Żyluk, 2018). Such a setting brings certain new dialogue moves into picture—we will encounter situations where the game master will ask the player directly to reformulate her/his question (see Section 2).

In what follows G refers to the player and M to the game master. Examples presented below are fragments of game sessions for puzzles "Drwale" (*Lumberjacks*), "Pragnienie" (*Thirst*), "Podróżnik" (*Traveler*), "Mecz" (*Football match*), "Kamienie" (*Pebbles*) and "Kosztowności" (*Valuables*). To give a better understanding of these examples I will provide now the main questions for these puzzles. The main problem for the *Lumberjacks* puzzle is the following: "Sailors cut down tops of masts. Why are they doing this?". For the *Thirst*: "One man lost his job when he quenched his thirst. Where was he employed?". For the *Traveler*: "A man, without a single visa, in one day visited eight different countries. Authorities of none of these countries tried to throw him out. What was his profession and how did he manage to do this?". For the *Football match*: "A radio commentator is giving a running commentary of a football match. He does not see players and he is making up their actions on the field on the fly. Why is he doing this?". For the *Pebbles*: "In an European city a man places pebbles around a building each night. What is this building and why this man is doing this?". And finally for the *Valuables*: "Today this object is treated only as a jewelry, however back in time it was used also for non-decorative purposes by certain people. Circumstances of such a use were not merry ones.".

As the ERC(TZ) contains transcriptions of a certain game between G and M we may expect to meet questions modifications similar to the ones observed in BNC and BEE. The motivation for these modification is probably analogous to the ones analysed in the context (3)—G wants to solve the presented puzzle s/he aims at getting the initial question resolved by M. As such s/he will modify and rephrase the initial question in order to facilitate the answering process. We observe open to open, but easier to answer question transitions, like (17) and (18).

Open to open, but more specific question (O–O)

(17) G: Czy to jest kwestia ciężaru statku? [*Is it somehow related to the weight of this ship?*]
  G: W sensie oni ucinają te czubki, żeby być lżejszym statkiem, żeby szybciej płynąć i szybciej uciec? [*I mean, are they cutting the tops of masts in order to make the ship lighter in order to escape?*]
  [*ERC(TZ):DrwaleB*24][8]
(18) G: Czy żeglarze planują opuścić statek? [*Are they planning to leave the ship?*]
  G: Czy próbują opuścić statek? [*Are they trying to leave the ship?*]
  [*ERC(TZ):DrwaleB*24]

Open to quasi-open (option suggested, O–QO)

We also encounter examples of transition from open to quasi-open question as in (19) and (20).

(19) G: A czy jest to jakaś osoba, która jakby wykonuje czynności dla pracowników ambasady? [*Is the case that this person is someone who does some kind of services for the embassy employees?*]

---

[8] This notation indicates ERC sub-corpus (TZ—the MindMaze game), the task name (Drwale) and the file number (B24). No sentence numbering is available for the ERC corpus. Original spelling is preserved in the examples. Translation from Polish to English by P.Ł.

G: Na przykład kierowca jakiś, ochroniarz, albo coś takiego? [*For example, some kind of driver, bodyguard or something like that?*]
[*ERC(TZ):PodroznikB4*]

(20) G: czy był to ktoś, powiedzmy, z kategorii medialnej? [*Was this someone from media?*]

G: W sensie, na przykład, dajmy na to, dziennikarz? [*I mean, let us say, a journalist, for example?*]
[*ERC(TZ):PodroznikB26*]

Open to yes/no question (O–Y/N)

Modification of an open $q1$ into a yes/no $q2$ may also be observed in the ETC(TZ) corpus data.

(21) G: Czy pogoda ma znaczenie? [*Is weather relevant?*]

G: Czy pogoda jest ładna, czy to jest spokojne morze? [*Is weather nice, do we have a calm sea?*]
[*ERC(TZ):DrwaleB24*]

Table 1 presents the summary of the presented and described question modification types. It is worth to point that the list of these types is certainly not exhaustive, as more corpus studies are required. What is visible is that in almost all examples (except of (15)) gathered for the needs of this paper $q1$ is an open question, which then require an easier to answer follow up question.

Table 1: The summary of the modification types described

Modification type	Example
Open to open (but more specific)	(3), (4), (5), (11), (12), (13), (14), (17), (18)
Open to quasi-open (option suggested)	(6), (7), (19), (20)
Open to choice (closed list of options)	(8), (9), (10)
Open to yes/no	(16), (21)
Choice to yes/no	(15)

2 Modification triggers

Up to this point we have observed examples where a move from $q1$ to $q2$ was triggered by a factor that is not visible in textual corpus data (see discussion in the Introduction). However, in BEE we can observe interesting cases of question modification which is somehow triggered by the answers provided by the second dialogue participant. This is understandable for the educational contexts. In these cases the initial question has to be reformulated in order to make it more understandable, easier to answer, or to ask about a related subject, which will in consequence lead to the answer to the initial question, or simply establish a common ground for dialogue participants. Such a modification is visible in the example (22) below.

(22) TUTOR: Do you remember reading in the lesson about the two different types of components across which you can take a voltage reading?
   STUDENT: no
   TUTOR: Do you remember reading about sources and loads?
   STUDENT: yes
   TUTOR: Great
   [BEE(F): Stud23]

In this case the modification of the initial question "Do you remember reading in the lesson about the two different types of components across which you can take a voltage reading?" is triggered by the student's answer "no". This creates the need of replacing the initial question with another one. Tutor asks whether the student remembers reading about sources and loads. This time the student provides an affirmative answer and the tutorial may be continued within the topic of loads and sources. Such a dialogue move is even more understandable when we consider a broader context, which leads to our example of the initial question rephrasing.

TUTOR: OK, let's think about this for a moment.
TUTOR: Do you see any potential problem with how you just hooked up the voltmeter leads?
STUDENT: yes
TUTOR: Good, what problem do you see?
STUDENT: they are both connected to the positive side.
TUTOR: Actually they are not.
TUTOR: What is considered the positive side or the negative side is relative to how you have the leads hooked up.
TUTOR: The positive side is everything between the red lead and the positive terminal on the battery.
TUTOR: The negative side is the rest of the circuit.
TUTOR: So you don't have a problem with polarity.
TUTOR: Can you identify another potential problem?
STUDENT: no
TUTOR: OK, I'll give you a hint then.
TUTOR: Do you remember reading in the lesson about the two different types of components across which you can take a voltage reading?
STUDENT: no
TUTOR: Do you remember reading about sources and loads?
   [BEE(F): Stud23]

The tutor's motivation behind dialogue moves leading to question modification is suggested by the fragment before this event. Tutor asks whether the student can identify a problem. After confronted with "no" answer, the tutor says that s/he will "give a hint" to a student in this situation. In what follows the tutor attempts to identify a topic which will be recognised by a student as known/understandable.[9]

Question modification triggered by the negative answer may be also found in the ERC(TZ) data. In (23) we observe how G's initial question is modified after M's answered "no" to the initial question ("embarked" vs. "trying to embark").

(23) G: Czy obcy żeglarze dostali się na pokład statku żeglarzy ścinających maszty? [*Is it the case that some other sailors embarked on the ship?*]
M: Nie. [*No*]
G: A czy próbują? [*Are they trying to do this?*]
[*ERC(TZ):DrwaleB*24]

And now let us take a closer look at even more interesting situations, i.e. when the game master (M) asks directly to reformulate the initial question. By this move s/he communicates *via* question a part of $q1$ which should be addressed in more details. What is especially interesting in examples below is that after the question is asked as a response to $q1$ (usually it is a clarification request) it does not lead to the answer, but to modifying $q1$ into yet another question—$q2$. What is more, $q2$ incorporates the information required by M. Thus we encounter three consecutive questions in such a dialogue.

(24) G: Czy budynek jest duży? [*Is this building big?*]
M: A jak byś zdefiniował „duży budynek"? [*How would you define a "big building"?*]
G: Czy to jest większe niż budynek mieszkalny? [*Is this building bigger that a living house?*]
[*ERC(TZ):Kamienie B*21]

In the example (24) above we observe the described scheme. G asks whether the building under discussion is big. As a query-response M asks a clarification question: "How would you define a 'big building'?". Instead of answering this question directly, G's dialogue move is to modify her/his initial question: "Is this building bigger that a living house?". New question asked by G incorporates the information called for by the game master (big building is a building bigger than a living house).

Similar behaviour may be observed in examples (25)–(27) below.

(25) G: Czy to jest budynek kulturalny? [*Is this building a cultural one?*]
M: Kulturalny... a w jakim sensie budynek kulturalny? [*Cultural one... in what sense it is a cultural building?*]

---

[9] More analysis of tutor's questioning strategies based on the BEE data may be found in (Łupkowski, 2017).

G: Związany z kulturą, historią, sztuką? Z kulturą? [*Related to culture, history, art? Related to culture?*]
M: Ale w jaki sposób definiujesz „związanie"? [*But, how would you define this „related"?*]
G: Związany... jest używany do celów kulturalnych, rozwojowych, dla ludzi. Edukacyjnych, w jakimś stopniu? [*Related... it is used for cultural purposes, development related issues, for people. To some extent educational ones?*]
[*ERC(TZ):Kamienie B21*]

(26) G: Czy te kamyki są duże? [*Are these rocks big?*]
M: A w jaki...? [*And how would you ...?*]
G: Większe niż pięść? [*Are they bigger than a fist?*]
[*ERC(TZ):Kamienie B21*]

(27) G: Chodzi o okoliczności śmierci? Używany był przy okoliczności czyjejś śmierci? [*Is it all about circumstances of death? Was its usage related to a circumstances of someone's death?*]
M: Jak byś zdefiniował okoliczność śmierci? W momencie? [*How would you define circumstances of someone's death? In moment of it?*]
G: Czy przedmiot ten był noszony w związku z żałobą po kimś? [*Was this object wore while mourning someone?*]
[*ERC(TZ):KosztownosciB22*]

In the above examples M used a clarification question of the form "how would you define". We may also point at the example of a simpler clarification request, like the one in (28) below.

(28) G: Dobra, to czy są to Niemcy? [*Right, are these Germans?*]
M: Komentujący? [*Commentators?*]
G: Nie, no, czy spotkanie odbywa się w Niemczech? [*No, i mean, is this meeting taking place in Germany?*]
[*ERC(TZ):Mecz B3*]

Examples presented in this section provide at least two possible triggering factors for question modification, which are visible on the textual level of a corpus data, i.e. the negative answer provided by a respondent and clarification question-response. I find the second one especially interesting, as we are dealing with situation when three consecutive questions are asked in a dialogue. What is more—as visible in (24)–(27)—clarification required by M is incorporated into $q2$ asked by G. I find it very intuitive in the light of the discussed rationale behind dialogical moves when we considering playing games (as discussed in the previous section). As G wants to solve the presented puzzle and for this gather information from M s/he will facilitate this process. That may be the reason why after M's clarification request G's reaction is not an explanation, but another question, which stays within the topic of $q1$, is more specific and incorporates the required clarifications—this seems an efficient strategy in the described context.

Summary

This paper presents an overview and preliminary analysis of dialogue moves of modifying or rephrasing initial question in information seeking dialogues. I provide natural language examples retrieved from three different corpora: BNC, BEE and ERC (ordered from the most general to the most specialised one). These examples are analysed in terms of different question modification types. The following modification types are noticed in the data: from open to open but more specific question; from open to quasi-open question (suggesting one of the options available); from open to choice question (where a list of options to choose from is offered); from open to yes/no question and from choice question to yes/no question. I am aware that this list is not complete as more types may be there in the corpus data. Thus more studies would be necessary, including systematic corpus study of BNC, BEE and ERC as well as other corpora offering dialogical data (like e.g., CHILDES, MapTask or SCORE). The aim will be to provide a typology of modifying moves with the respect of its role for a dialogue analogical to the one offered by Łupkowski and Ginzburg (2016) for query-responses and for question response space by Ginzburg et al (2019). Majority of the gathered examples stays in line with the erotetic decomposition principle described in the Introduction. As such they may be analysed with the use of formal tools including Inferential Erotetic Logic (Wiśniewski, 2013), (Łupkowski et al, 2018a), KoS (Ginzburg, 2012) and formal dialogue systems (Łupkowski, 2016). Such modelling is planned in future research on this subject.

I also discuss the issue of potential triggers of the initial question modification. This covers triggers which are not present in textual corpus data, but also these which may be observed there (including a negative answer provided by other dialogue participant, as well as clarification request formulated by this participant). As mentioned in the Introduction, this aspect of questions modification requires further studies. I find corpora aligned with dialogue recordings especially promising in this context (like e.g. ERC(TZ) and SCORE). I hope that examples gathered for the purposes of this paper and their preliminary analysis offer a good starting point and ramification for the future studies.

References

Burnard L (ed) (2007) Reference guide for the British National Corpus (XML Edition). Oxford University Computing Services on behalf of the BNC Consortium, URL http://www.natcorp.ox.ac.uk/XMLedition/URG/, acess 20.03.2017

Ginzburg J (2012) The interactive stance: Meaning for conversation. Oxford University Press, Oxford

Ginzburg J, Yusupujiang Z, Li C, Ren K, Łupkowski P (2019) Characterizing the response space of questions: a corpus study for english and polish. In: Proceedings of the 20th Annual SIGdial Meeting on Discourse and Dialogue, pp 320–330

Kuppevelt JV (1995) Discourse structure, topicality and questioning. Journal of Linguistics 31:109–147

Marciniak B, Łupkowski P (2017) Erotetic Reasoning Corpus Tools: Search & Browse Tool. Tech. rep., Adam Mickiewicz University, Poznań, uRL=https://ercorpus.wordpress.com/tools/

Purver M (2001) SCoRE: A tool for searching the BNC. Tech. Rep. TR-01-07, Department of Computer Science, King's College London, URL `ftp://ftp.dcs.kcl.ac.uk/pub/tech-reports/tr01-07.ps.gz`

Rosé CP, Eugenio BD, Moore JD (1999) A dialogue-based tutoring system for basic electricity and electronics. In: Lajoie SP, Vivet M (eds) Artificial intelligence in education, IOS, Amsterdam, pp 759–761

Urbański M, Paluszkiewicz K, Urbańska J (2016) Erotetic problem solving: From real data to formal models. an analysis of solutions to erotetic reasoning test task. In: Paglieri F, Bonetti L, Fellett S (eds) The Psychology of Argument: Cognitive Approaches to Argumentation and Persuasion, College Publications, London, pp 33–46

Urbański M, Żyluk N (2018) Sets of situations, topics, and question relevance. In: Oswald S (ed) Proceedings of the 2nd European Conference on Argumentation (to appear), College Publications, London

Wiśniewski A (2013) Questions, inferences and scenarios. College Publications, London

Łupkowski P (2016) Logic of Questions in the Wild. Inferential Erotetic Logic in Information Seeking Dialogue Modelling. College Publications, London

Łupkowski P (2017) IEL-based formal dialogue system for tutorials. Logic and Logical Philosophy 26(3):287–320

Łupkowski P, Ginzburg J (2013) A corpus-based taxonomy of question responses. In: Proceedings of the 10th International Conference on Computational Semantics (IWCS 2013), Association for Computational Linguistics, Potsdam, Germany, pp 354–361

Łupkowski P, Ginzburg J (2016) Query responses. Journal of Language Modelling 4(2):245–293

Łupkowski P, Leszczyńska-Jasion D (2014) Generating cooperative question-responses by means of erotetic search scenarios. Logic and Logical Philosophy 24(1):61–78

Łupkowski P, Majer O, Peliš M, Urbański M (2018a) Epistemic erotetic search scenarios. Logic and Logical Philosophy 27(3):301–328

Łupkowski P, Urbański M, Wiśniewski A, Błądek W, Juska A, Kostrzewa A, Pankow D, Paluszkiewicz K, Ignaszak O, Urbańska J, et al (2018b) Erotetic reasoning corpus. a data set for research on natural question processing. Journal of Language Modelling 5(3):607–631

Interleaved Argumentation and Explanation in Dialog

Adrian Groza

Abstract Computational models for natural arguments are more realistic when they encompass concepts of both argumentation and explanation, as shown in the informal logic literature. Apart from distinguishing explanations from arguments, I am presenting our approach for modeling them in description logic. By using description logics (DL) to define the ontologies of the agents, the DL reasoning tasks are used to distinguish an argument from an explanation.

Key words: argumentative agents, explanatory agents, reasoning in description logic, explainable AI

1 Introduction

Argument and explanation are considered distinct and equally fundamental (Mayes, 2000), with a complementary relationship (Mayes, 2010; Arioua et al, 2017; Bex and Walton, 2016), as a central issue for identifying the structure of natural dialogs. While argumentation brings practical benefits in persuasion, deliberation, negotiation, collaborative decisions, or learning (Xu et al, 2020; Guid et al, 2019), it also involves costs (Paglieri and Castelfranchi, 2010). Differently, the complementary domain of explanation (Miller, 2018) has not met the same level of formalization as argumenation (Pearl and Mackenzie, 2018). However, formalizing explanations could benefit from the recent work under the umbrella of explainable artificial intelligence (XAI) (Gunning, 2017).

We aim here to distinguish between argument and explanation in natural dialog. Even if interleaving argument and explanation is common practice in daily commu-

---

Adrian Groza
Department of Computer Science
Technical University of Cluj-Napoca, Romania
e-mail: Adrian.Groza@cs.utcluj.ro

nication, the task of extending argumentation theory with the concept of explanation is still at the early stages. Given the interleaving of arguments and explanations in natural dialog, we are interested here in modeling arguments and explanations in Description Logic (DL). By exploiting the reasoning tasks of the DL, the system we implemented is able to automatically classify arguments and explanations, based on the partial information disclosed during dialog. To facilitate situation awareness and common understanding during dialog, we also model subjective perspective of agents on arguments and explanation. The main benefit of our Argument-Explanation ontology is that agents can identify more quickly agreements and disagreements during dialogs. By early signaling misunderstandings, the agents will avoid conveying speech acts that are inadequate for the current state of the dialog.

The fusion of argument and explanation is best shown by the fact that humans tend to make decisions based both on knowledge and understanding (Wright, 2002). For instance, in judicial cases, circumstantial evidence needs to be complemented by a motive explaining the crime, but the explanation itself is not enough without plausible evidence (Mayes, 2010). In both situations the pleading is considered incomplete if either argumentation or explanation is missing. Thus, the interaction between argument and explanation, known as *argument-explanation pattern*, has been recognized as the basic mechanism for augmenting an agent's knowledge and understanding (de Vries et al, 2002).

The relation between knowledge, argument, and explanation is also covered in this study. Firstly, our starting point is the role of knowledge in argumentation, as stressed out by Walton and Godden (2007). In natural dialog knowledge is interleaved with argumentation. For instance, when performing reasoning tasks on available knowledge, agents perform better if the reason is argumentative (Mercier and Sperber, 2011). On the one hand, knowledge of agents is exploited when generating, conveying, and assessing arguments (Walton and Godden, 2007). On the other hand, argumentation can be an efficient tool for knowledge acquisition (Amgoud and Serrurier, 2008) or collaborative knowledge engineering (Tempich et al, 2005). Secondly, explanation aims to transfer understanding. For human agents, understanding occurs in different degrees, relative to their knowledge bases, beliefs, and goals. Cognitive understanding requires similar ontologies, but assumes that agents have different goals and beliefs.

This work offers a precise distinction between argument and explanation in a dialogue, and models it in Description Logics. Preliminary ideas of this paper have been discussed at the CMNA workshop (Letia and Groza, 2012) and Poznań Reasoning Week (Groza, 2018).

2 Distinguishing Argument from Explanation

The role of argument is to establish knowledge, while the role of explanation is to facilitate understanding (Mayes, 2010). Thus, to make an instrumental distinction between argument and explanation, one has to distinguish between knowledge and understanding. One legitimate question would be: does understanding represent more knowledge? By defining both concepts in terms of the epistemic notion of awareness, knowledge represents awareness of information, while understanding represents the awareness of the relations between items of information. Thus, understanding is a form of organization of justified beliefs (Janvid, 2012). In the simplest computational model, understanding of a concept can be quantified in terms of the number of relations an agent is aware of in a given context regarding that concept. A supplementary constraint would impose these relations to include causal, and other types of roles among them, in order to assign a meaning to the concept. From an operational or behavioral viewpoint, understanding allows the knowledge to be put into practice. On this line, understanding represents a deeper level than knowledge.

We restricted ourselves here to a causal model for explanation (Pearl and Mackenzie, 2018). This restriction is justified by two operational objectives: First, we want to build a formal model of arguments and explanation. The restriction facilitates the formalization of distinguishing features of argument and causal explanation in Description Logic. Second, we consider explanation in the context of dialogues, and causal explanations are seen as a form of social interaction, stated by Hilton as:

> Causal explanation is first and foremost a form of social interaction. One speaks of giving causal explanations, but not attributions, perceptions, comprehensions, categorizations, or memories. [...] Causal explanation takes the form of conversation and is thus subject to the rules of conversation. (Hilton, 1990)

We consider the following distinctive features of argument and explanation:

Starting condition. Explanation starts with non-understanding. Argumentation starts with a conflict.
Role symmetry. In explanation the roles are usually asymmetric: the explainer is assumed to have more understanding and wants to transfer it to the explainee. In argumentation, both parties start the debate from equal positions, thus initially having the same roles. Only at the end of the debate the asymmetry arises when the winner is considered to have more relevant knowledge on the subject. If no winner occurs, the initial symmetry between arguers is preserved.
Linguistic indicator. In explanation one party supplies information. There is a linguistic indicator which requests that information. Because in argumentation it is assumed that all parties supply information, no indicator of demanding the information is required.
Acceptance. An argument is accepted or not, while an explanation may have levels of acceptance.

Regarding the "starting condition", for an argument, premises represent evidence supporting a doubted conclusion. For an explanation, the conclusion is accepted and the premises represent the causes of the consequent (see Fig. 1). The explanation aims to understanding the explanandum by indicating what causes it, while an argument aims to persuade the other party about a believed state of the world. An argument is considered adequate if there is at least one agent who justifiably believes that the premises are true but who does not justifiably believe this about the consequent (Lumer, 2005). An

Fig. 1: Distinguishing argument from explanation

explanation is adequate if all the agents accepting the premises would also accept the consequent. The function of argument is to "transfer a justified belief", while the role of explanation is to "transmit understanding". Therefore, unlike arguments, the statements in an explanation link well known consequents to less known premises (Hempel and Oppenheim, 1948).

Regarding the "role symmetry", consider a dialog between a teacher and a junior student which is almost entirely explicative. The ontology of the student regarding the specific scientific field is included in the ontology of the teacher. As the ontology of the student increases, resulting in different perspectives on the subject, exchanging arguments may occur. The above teacher-student scenario helps us to extract several knowledge conditions for arguments. A doubted conclusion may arise from the differences in the knowledge bases of the two agents. Assuming the same reasoning capabilities, the precondition would be for the agents to have different ontologies for arguments to arise. Formally, the intersection between agents ontologies shouldn't be empty ($\mathcal{O}_i \cap \mathcal{O}_j = \mathcal{O}_{ij} \neq \varnothing$), so that the agents can communicate, but the differences should be substantial enough to generate arguments. The arguments are constructed based on knowledge in the symmetric difference of the agents ontology $\mathcal{O}_i \Delta \mathcal{O}_j = \mathcal{O}_i \setminus \mathcal{O}_j \cup \mathcal{O}_j \setminus \mathcal{O}_i$. Depending on the granularity of the common ontology $\mathcal{O}_{ij}$, one agent would convey more abstract or more concrete arguments in order to adapt them to the audience.

For "linguistic indicator", a mean to distinguish between explanation and argument is to compare arguments *for F* and explanations *of F*. The mechanism should distinguish between whether F is true and why F is true. In case F is a normative sentence, the distinction is difficult (Wright, 2002). If F is an event, the question why F happened is clearly delimited by whether F happened.

The "acceptance" topic is supported by the fact that, unlike knowledge, understanding admits degrees (Janvid, 2012). The smallest degree of understanding, making sense, demands a coherent explanation, which usually is also an incomplete one. It means that, when the explainer conveys an "I understand" speech act, the explainer can shift to an examination dialog in order to figure out the level of understanding, rather than a crisp value understand/not understand, as investigated by Walton (2011). Acceptability standards for evaluating explanation can be defined similarly to standards of proof in argumentative theory (Gordon et al, 2007). The elements used to distinguish between argument and explanation are collected in Table 1.

Table 1: Explanations versus arguments

	Explanation	Argument
Consequent	Accepted as a fact	Disputed by parties
Premises	Represent causes	Represent evidence
Reasoning Pattern	Provides less well known statements why a better known statement is true	From well known statements to statements less well known
Answer to	Why is that so?	How do you know?
Contribute to	Understanding	Knowledge
Acceptance	Levels of understanding	Yes/No

3 Representation for Argument and Explanation

3.1 Description Logics

In description logic (DL) there are concepts (C) and relations (or roles r) among these concepts. Roles are quantified universally ($\forall r.C$), existentially ($\exists r.C$), or explicitly stating the number of roles pointing towards the specific concepts (e.g. $(=1)r.C$). Axioms for concepts and roles are stored in a terminological box (*TBox*). To indicate that the individual i is an instance of the concept C, the notation $i{:}C$ is used. The expression $(i,j){:}r$ says that the individuals i and j are related by the role r. The set of all individuals are shown in the assertional box (*ABox*).

3.2 Arguments and Explanations in Description Logic

At the top level of our argument and explanation ontology (*ArgExp*), we have *statements* and *reasons*. A statement claims a text of type string, given by:

$$Statement \sqsubseteq \exists\, claimsText.String.$$

Definition 1. A reason consists of a set of premises supporting one conclusion.

$$Reason \sqsubseteq \exists\, hasPremise.Statement \sqcap (=1)hasConclusion.Statement \quad (1)$$

Arguments and explanations are forms of reasoning.

Definition 2. An argument is a reason in which the premises represent evidence in support of a doubted conclusion.

$$Argument \sqsubseteq Reason \sqcap \forall\, hasPremise.Evidence \sqcap (=1)hasConclusion.DoubtedSt \quad (2)$$

Definition 3. An explanation is a reason in which the premises represent a cause of an accepted fact.

$$Explanation \sqsubseteq Reason \sqcap \forall\, hasPremise.Cause \sqcap (=1)hasConclusion.Fact \quad (3)$$

We define a doubted statement as a statement attacked by another statement:

$$DoubtedSt \equiv Statement \sqcap \exists\, attackedBy.Statement \qquad (4)$$

The domain of the role *attackedBy* is a *Statement* ($\exists\, attackedBy.\top \sqsubseteq Statement$), while its range is the same concept *Statement*: $\top \sqsubseteq \forall\, attackedBy.Statement$. The role *attacked* is the inverse role for *attackedBy*, expressed in DL with $attack^- \equiv attackedBy$.

A fact is a statement which is not doubted.

$$Fact \equiv Statement \sqcap \neg DoubtedSt \qquad (5)$$

Note that facts and doubted statements are disjoint ($Fact \sqcap DoubtedStatement \sqsubseteq \bot$). Pieces of evidence and cause represent statements.

$$Evidence \sqsubseteq Statement \qquad (6)$$
$$Cause \sqsubseteq Statement \qquad (7)$$

The concepts for evidence and cause are not disjoint: the same sentence can be interpreted as evidence in one reason and as cause in another reason, as illustrated in Example 1.

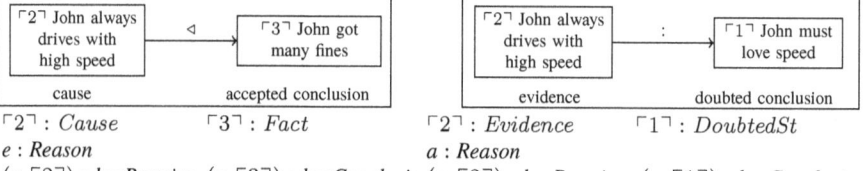

Fig. 2: The same statement ⌜2⌝ acts as a cause for the accepted statement ⌜3⌝ and as evidence for doubted statement ⌜1⌝. The agent with this interpretation function treats e as an explanation ($e : Explanation$) and a as an argument ($a : Argument$)

Example 1 (Different interpretations of the same premise.). Consider the following statements:

> John must love speed. ⌜1⌝
> He drives with high speed all the time. ⌜2⌝
> That's why he got so many fines. ⌜3⌝

One possible interpretation is that statement ⌜2⌝ represents the support for statement ⌜1⌝. Statement ⌜2⌝ also acts as an explanation for ⌜3⌝, as suggested by the textual indicator *"That's why"*. Fig. 2 illustrates the fomalisation in DL of these two reasons. Assume that the interpretation function $\mathcal{I}$ of the hearing agent h asserts statement ⌜2⌝

Interleaved Argumentation and Explanation in Dialog 121

as an instance of the concept *Cause* and ⌜3⌝ as a *Fact*. Based on axiom 3, agent h classifies the reason e as an explanation.

Assume that Abox of agent h contains also the assertion (⌜1'⌝,⌜1⌝):*attacks*. Based on axiom 4, agent h classifies the statement ⌜1⌝ as doubted. Adding that ⌜2⌝) is interpreted as evidence, agent h classifies the reason a, based on definition 2. The relations among individuals in the Example 1 are depicted in Fig. 3. Here, the top level concepts of our argument-explanation ontology *ArgExp* are also illustrated. Based on the definitions in the *TBox* and the instances of the *ABox*, a is an argument and e is an explanation.

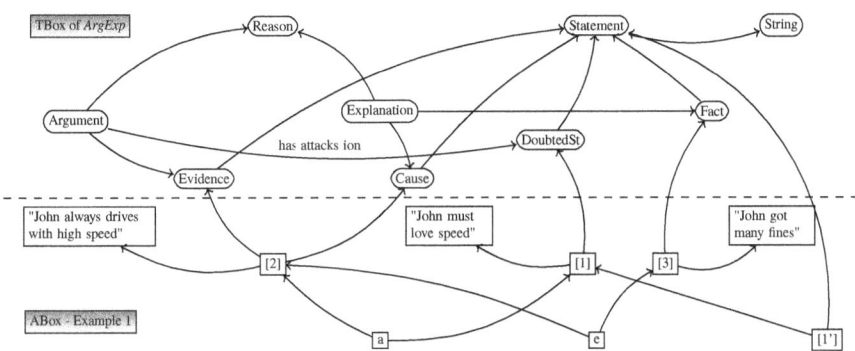

Fig. 3: Graphical representation of the Tbox and Abox of the agent h regarding Example 1

Agents can have different interpretation functions of the same chain of conveyed statements. In Example 2, the agents have opposite interpretation regarding the premise and the conclusion of the same reason.

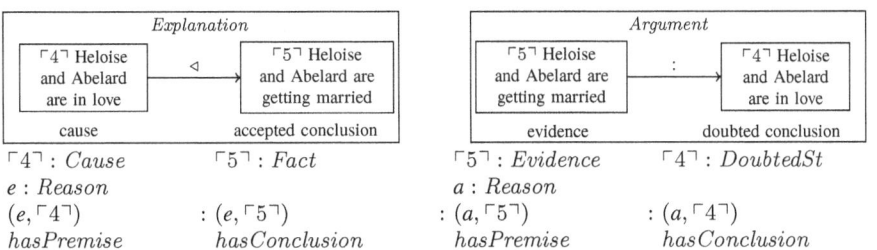

Fig. 4: Opposite interpretations of the same reason: In the left part, e is classified as an explanation ($e : Explanation$). In the right part, a is interpreted as an argument ($a : Argument$)

Example 2 (Opposite interpretations of the same reason.). Consider the following reason containing the statements ⌜4⌝ and ⌜5⌝:

> *Heloise and Abelard are in love.*   ⌜4⌝
> *Heloise and Abelard are getting married.* ⌜5⌝

The ambiguity arises from the difficulty to identify which is the premise and which the conclusion. One agent can interpret ⌜4⌝ as a cause for the accepted fact ⌜5⌝, treating the reason *e* as an explanation (left part of Fig. 4). Here, ⌜4⌝ acts as a premise in the first interpretation (left part) and as a conclusion in the second one (right part). An agent with a different interpretation function asserts ⌜5⌝ as evidence for the doubted conclusion ⌜4⌝, therefore rising an argument.

To remove the ambiguity, agents can exploit the information that the given dialog is interpreted as an explanation by one party and as an argument by the other. Consider the following dialog adapted from (Budzynska and Reed, 2011):

Bob: *The government will inevitably lower the tax rate.* ⌜6⌝
Wilma: *How do you know?*   ⌜7⌝
Bob: *Because lower taxes stimulate the economy.*   ⌜8⌝

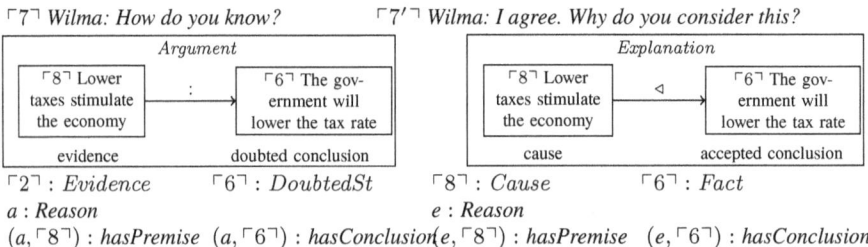

Fig. 5: The dialog provides indicators helping Bob to assess the status of the consequent from Wilma's perspective: In the left part, query ⌜7⌝ does not suggest the acceptance of conclusion ⌜6⌝. In the right part, answer ⌜7'⌝ clearly indicates the Wilma accepts claim ⌜6⌝

The dialog is shown in the Fig. 5 as an argument with the consequent ⌜6⌝ supported by the premise ⌜8⌝. Let's assume that Wilma's reply is slightly modified, given by:

⌜7'⌝ Wilma: *I agree. Why do you consider this?*

By accepting statement ⌜6⌝, it becomes a fact in the situation represented by Bob and Wilma. Consequently, the reason becomes an explanation in which the cause "lower taxes stimulate the economy" may explain the government's decision (Fig. 5). Under the assumption that an agent accepts a statement only if it has a level of understanding of that sentence, one can infer that Wilma has her own explanation regarding the fact ⌜6⌝, but she wants to find out her partner's explanation.

Another issue regards the distinction between evidence and cause. Cognitive experiments (Brem and Rips, 2000) have shown difficulties when distinguishing between them, where only 74% of the subjects have correctly classified pieces of information as evidence or cause. Moreover, human agents are able to build a strategy of substituting explanation in the case that the evidence is not available (Brem and Rips, 2000). Given the difficulty to distinguish between causes and evidence, a simplified argument-explanation model would consider only the status of the consequent. Thus, if an agent accepts the conclusion according to its interpretation function, then it treats the premise as cause (axiom 8). If the agent interprets the conclusion as doubted, it will treat the premise as evidence (axiom 9).

$$\exists hasPremise^-.(Reason \sqcap \exists hasConclusion.Fact) \sqsubseteq Cause \quad (8)$$
$$\exists hasPremise^-.(Reason \sqcap \exists hasConclusion.DoubtedSt) \sqsubseteq Evidence \quad (9)$$

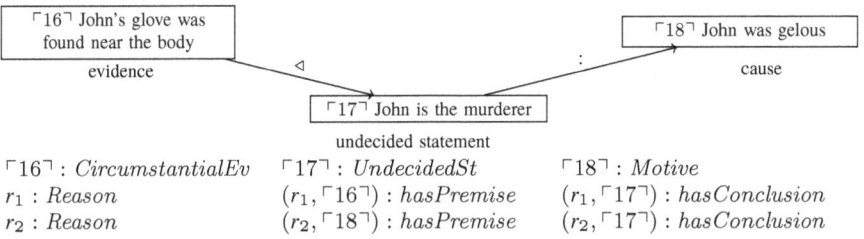

⌜16⌝ : $CircumstantialEv$
r_1 : $Reason$
r_2 : $Reason$

⌜17⌝ : $UndecidedSt$
$(r_1, ⌜16⌝)$: $hasPremise$
$(r_2, ⌜18⌝)$: $hasPremise$

⌜18⌝ : $Motive$
$(r_1, ⌜17⌝)$: $hasConclusion$
$(r_2, ⌜17⌝)$: $hasConclusion$

Fig. 6: Argument-explanation pattern supporting consequent

3.3 Argument-explanation pattern

In many situations, people use both evidence and explanations to complementarily support the same consequent. Many examples come from law. Lawyers start their pledge by using the available evidence to persuade the jury about a claim which is not assumed accepted. When the jury tend to accept the claim, the lawyer provides explanations why the event took place as it really happened.

An argument-explanation pattern occurs in two steps:

1. In the first step, evidence e is provided for supporting claim s, with s assumed undecided at the current moment,
2. In the second step, cause c is used to explain why the same statement s took place, with s assumed plausibly accepted by the audience in the light of previous evidence e (example 3).

Example 3 (Argument-explanation pattern). ⌜16⌝ John's glove was found near the body of his wife's friend. ⌜17⌝ John is the murderer. He committed the murder because ⌜18⌝ he was jealous.

To accommodate the argument-explanation pattern in the Fig. 6, we firstly need to introduce the concept of undecided statement $UndecidedSt$, disjoint with a doubted or an accepted statement.

$$UndecidedSt \sqsubseteq Statement \qquad (10)$$
$$UndecidedSt \sqsubseteq \neg DoubtedSt \qquad (11)$$
$$UndecidedSt \sqsubseteq \neg Fact \qquad (12)$$

Secondly, we refined the *ArgExp* ontology by classifying evidence in direct or circumstantial, depending on the type of support provided for it.

$$DirectEv \sqsubseteq Evidence \sqcap \exists\, directsup.DoubtedSt \qquad (13)$$
$$CircumstantialEv \sqsubseteq Evidence \sqcap \exists\, indirectsup.DoubtedSt \qquad (14)$$

A motive is a more specific cause, $Motive \sqsubseteq Cause$.

To formalise the argument-explanation pattern exemplified above, we need rules on top of *ArgExp*:

Definition 4. An argument-explanation pattern is a tuple $\langle e, c, u \rangle$ with e interpreted as evidence, c as cause, and u as undecided statement, constructed by the rule:

$$\begin{aligned}\langle e,c,u\rangle \Leftarrow\ & e: Evidence \wedge c: Cause \wedge u: UndecidedSt \wedge\\ & pa: PossibleArg \wedge pe: PossibleExp \wedge\\ & (pa,e): hasPremise \wedge (pa,u): hasConclusion\\ & (pe,c): hasPremise \wedge (pe,u): hasConclusion\end{aligned} \qquad (15)$$

where

$$\begin{aligned}PossibleArg \sqsubseteq\, & Reason \sqcap \forall\, hasPremise.Evidence\\ & \sqcap (=1) hasConclusion.UndecidedSt\end{aligned} \qquad (16)$$

$$\begin{aligned}PossibleExp \sqsubseteq\, & Reason \sqcap \forall\, hasPremise.Cause \sqcap\\ & (=1) hasConclusion.UndecidedSt\end{aligned} \qquad (17)$$

Interplay between arguments and explanations can lead to more complex reasoning patterns, such as an explanation followed by an argument or an argument followed by an explanation. In the first case, the doubted conclusion of the argument is used as a cause for an explanation. In the second case, the fact supported by an explanation is used as evidence for the premise of an argument.

Agent A ($\mathcal{O}_A$)
$u : GoodUniversity$
$GoodUniversity \sqsubseteq$
$\exists hasGood.ResearchFacility$

Agent's A view on agent B ($\mathcal{O}_{AB}$)
$u : GoodUniversity$
$GoodUniversity \equiv \exists hasGood.ResearchFacility \sqcap$
$\exists hasGood.TeachingFacility$

Agent B view on agent A ($\mathcal{O}_{BA}$)
$u : ResearchInstitute$
$ResearchInstitute \sqsubseteq$
$\exists hasGood.ResearchFacility$

Agent B ($\mathcal{O}_B$)
$u : GoodUniversity$
$GoodUniversity \equiv \exists hasGood.ResearchFacility \sqcup$
$\forall hasGood.TeachingFacility$

Fig. 7: Subjective views of agents

4 The Subjective Views of the Agents

The agents construct arguments and explanations from their own knowledge bases which do no completely overlap. At the same time, each party has a subjective model about the knowledge of its partner.

Let's consider the partial knowledge in Fig. 7. Here, agent *A* sees the individual *u* as a good university, where a good university is something included in all objects for which the role *hasGood* points towards concepts of type *ResearchFacility*. According to the agent *B* ontology ($\mathcal{O}_B$), *u* is also a good university, but the definition is more relaxed: something is a good university if it has at least one good research facility or all the teaching facilities are good.

According to the agent *A*'s perspective on the knowledge of the agent *B* ($\mathcal{O}_{AB}$), *u* belongs to the concept of good universities, but the definition is perceived as being more restrictive: a good university should have at least one good research facility but also at least one good teaching facility. From the opposite side ($\mathcal{O}_{BA}$), the agent *B* perceives that *A* asserts *u* as a research institute, where a research institute should have good research facility.

Suppose that the agent *A* conveys different reasons s_1 and s_2 supporting the statement c_1: "u has good research facility" and c_2: "u has either good research or good teaching". For instance:

s_1: "Because *u* attracted large funding from research projects, *u* manages to build a good research facility."

s_2: "Because *u* attracted large funding from research projects, *u* should have either good research or good teaching."

The reasons s_1 and s_2 are graphically represented in the Fig. 8. Let's assume that both agents formalize statements c_1 and c_2 as follows:

c_1 : "$u : \exists hasGood.ResearchFacility$"
c_2 : "$u : \exists hasGood.(ResearchFacility \sqcup Teaching)$"

How does the agent *A* treat one reason, when conveying it to the agent *B*, as explanation or argument?. Given the models in the Fig. 7, how does the receiving agent *B* perceive the reason: an explanatory or a persuasive one?

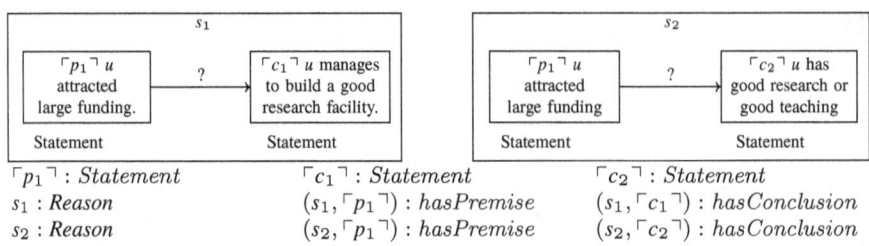

⌜p_1⌝ : Statement
s_1 : Reason
s_2 : Reason

⌜c_1⌝ : Statement
(s_1, ⌜p_1⌝) : hasPremise
(s_2, ⌜p_1⌝) : hasPremise

⌜c_2⌝ : Statement
(s_1, ⌜c_1⌝) : hasConclusion
(s_2, ⌜c_2⌝) : hasConclusion

Fig. 8: Possible reasons conveyed by the agent A. Are they arguments or explanations?

To distinguish between explanation and argument, the most important issue regards the acceptance of the consequent. In Table 2, $\oplus$ denotes that the ontology $\mathcal{O}_X$ entails the consequent c_j. The statement c_1 can be derived from the ontology $\mathcal{O}_A$ (Fig. 7). It cannot be inferred by the agent B based on its ontology $\mathcal{O}_B$ (noted with $\ominus$). That is because B considers a university which has only good teaching facilities, but no good research facility, is also a good university (given by the disjunction in the definition of *GoodUniversity* in $\mathcal{O}_B$).

Table 2: Entailment of statements c_1 and c_2 in agent ontology

Agents ontologies $\models c_1$?	$\models c_2$?
$\mathcal{O}_A$	$\oplus$ $\oplus$
$\mathcal{O}_{AB}$	$\oplus$ $\ominus$
$\mathcal{O}_B$	$\ominus$ $\oplus$
$\mathcal{O}_{BA}$	$\oplus$ $\oplus$

Instead, the statement c_2 fits the definition of good ontology in $\mathcal{O}_B$. Because the agent A accepts its first part *"u has good research"*, it should also consider c_2: *"u has good research or good teaching"* as valid. Similarly, the agent A considers that the agent B cannot infer c_2 ($\ominus$ in the Table 2), even if the $\mathcal{O}_B$ ontology entails c_2.

The agent A has a wrong representation $\mathcal{O}_{AB}$ regarding how the agent B views the statement c_2. Even if the agent B has a wrong model $\mathcal{O}_{BA}$, based on which it believes that the agent A interprets u as a research institute instead of a university, the consequent c_2 is still derived based on the axiom

$$ResearchInstitute \sqsubseteq \exists hasGood.ResearchFacility.$$

The knowledge of the agent A ($\mathcal{O}_A$), and its model about the knowledge of B ($\mathcal{O}_{AB}$), represent the subjective world of the agent A, noted with w_A in the Table 3. Similarly, the subjective world w_B of the agent B consists of the knowledge of B ($\mathcal{O}_B$), and its view on the knowledge of the agent A. The knowledge of A combined with the knowledge of B ($\mathcal{O}_{BA}$), represent the objective world w_O. A statement is considered *Accepted* if it

Table 3: Acceptance of consequents c_1 and c_2 based on ontology

World Ontologies	c_2	c_1
w_O	$\mathcal{O}_A + \mathcal{O}_B$	Accepted Doubted
w_A	$\mathcal{O}_A + \mathcal{O}_{AB}$	Doubted Accepted
w_B	$\mathcal{O}_B + \mathcal{O}_{BA}$	Accepted Doubted

is entailed by both ontologies. If at least one ontology does not support the statement, it is considered *Doubted*. The following algebra encapsulates this:

$$\oplus + \oplus = Accepted \quad \oplus + \ominus = Doubted$$
$$\ominus + \oplus = Doubted \quad \ominus + \ominus = Doubted$$

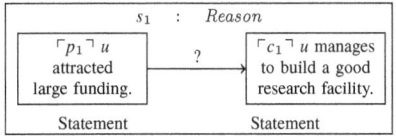
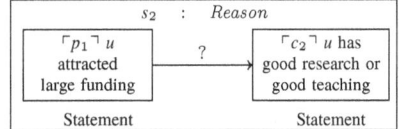

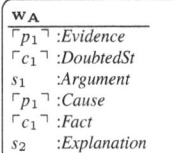
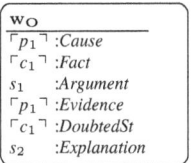
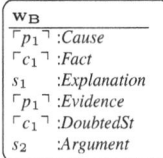

Fig. 9: Interpreting reasons s_1 and s_2 in different worlds

In the Table 3, the agent A treats c_2 as accepted, meaning that from its point of view the reason s_2 represents an explanation. The agent B perceives the sentence c_2 as doubted, therefore it considers that it is hearing an argument (world w_A in the Fig. 9). Note that in the objective world w_O, the reason s_2 is actually an argument. That means that the agent A is wrong about the model of its partner B. Consider that the reason s_1 is uttered by the agent B. It believes that it is conveying an argument, which is true in the objective world w_O. The agent A considers that it is receiving an explanation, which is false in w_O.

The statement c_1 being perceived as doubted in w_A, the agent A considers that it is conveying an argument. In the world w_B, the conclusion is accepted, thus the agent B is hearing an explanation, which is true in the objective world w_O. In this situation, the agent B should signal to its partner: "There is no need to persuade me. I agree with the consequent."

The correctness or adequacy of conveying either argument or explanation should be computed relative to the objective world w_O. Given the difference between expecting explanations or arguments (subjective worlds w_A and w_B) and legitimate ones (objective world w_O), the agents may wrongly expect explanations instead of arguments and *vice versa*. For the correctness or adequacy of conveying/expecting argument or explanation, the algebra in the Table 5 is used. The first operator represents the actual world w_O, while the second one represents the subjective perspective of the agent X.

Table 4: Cases when X conveys/expects argument or explanation

	Communicate	Expects
Argument	$Doubted_X$	$\oplus_X^w \vee \ominus_X^{\neg w}$
Explanation	$Doubted_X$	$\ominus_X^w \vee \oplus_X^{\neg w}$

By analyzing the entailment of a statement in all four knowledge bases, the situations in which the agents expect explanation or argument are synthesized in the Table 4. Assuming sincere agents, X conveys an argument if in its world the statement is *Doubted*. If the statement is *Accepted*, X conveys explanation. The agent X receives explanations when it is right about an agreement ($\oplus_X^w$) or when it is not aware of a conflict ($\ominus_X^{\neg w}$). It receives arguments either when X is aware of a disagreement ($\ominus_X^w$) or it is not aware of an agreement ($\oplus_X^{\neg w}$).

Table 5: Correctness/inadvertence of expectation

$Accepted_O + Accepted_X = \oplus_X^w$ agreement rightness
$Accepted_O + Doubted_X = \oplus_X^{\neg w}$ agreement not aware
$Doubted_O + Accepted_X = \ominus_X^{\neg w}$ conflict not aware
$Doubted_O + Doubted_X = \ominus_X^w$ conflict rightness

The situation resulting by applying the algebra in the Table 5 on the given scenario is presented in the Table 6. The agent B, even if its model about A is not accurate, manages to figure out the status of both consequents c_1 and c_2. Quite differently, the agent A is ignorant with respect to both conclusions.

Table 6: Awareness regarding consequents c_1 and c_2

Agent	Awareness and Ignorance	c_1	c_2
A	$w_O + w_A$	$\ominus_A^{\neg w}$	$\oplus_A^{\neg w}$
B	$w_O + w_B$	$\ominus_B^w$	$\oplus_B^w$

Is it possible for the hearing agent to indicate to the conveyor agent that a wrong assumption has been made? The problem is that no agent is aware of the objective world w_O. At least two options may exist to solve this issue:

1. If a mediator would exist, aware of w_O, it would be able to identify misunderstandings and to provide guidance for increasing the dialog efficiency.
2. The second option would be to introduce distinctive communicative acts for conveying either explanation or argument. The consequence is minimizing misunderstandings in dialog, because agents can better understand the cognitive maps of their partners.

For instance, if the agent X announces that s_1 is an explanation, its partner Y can disclose instantly its doubts about the conclusion of s_1. By updating its model $\mathcal{O}_{XY}$, the agent X can re-interpret s_1 as an argument, at this specific moment of the conversation. Thus, incorrect assumptions about accepted or doubted statements are eliminated as soon as they explicitly appear in the dialog. Moreover, people do use this kind of distinction in their discourses, when framing their speech with: "I'll try to explain for you", "One explanation is...", "The main cause is", "My argument is..." etc. These distinctive speech acts for conveying argument or explanation do support better coomunication among agents. The decision when to use an argumentative speech act or an explanatory one is based on reasoning on the proposed ArgExp ontology.

5 Discussion and Related Work

Joined argument and explanation.

Argumentation and explanation have been combined in computational models, starting with Shanahan (1989) and Poole (1989). Bex et al (2010) have exploited the argument-explanation complementarity for legal reasoning, while (Moulin et al, 2002) for building more persuasive agents. Interleaving argument and explanation in natural dialog has been investigated in (Bex and Prakken, 2008) and (McBurney and Parsons, 2001). Zeng et al. have proposed an argumentation-based approach for making context-based and explainable decisions (Zeng et al, 2018). Two explanatory patterns have been formalized: *argument-explanation* and *context-explanation*. Zeng et al. have focused on context in a single agent setting, whereas we did not focus on context formalization, but on subjective views of multi-agents. Except for McBurney and Parsons (2001), the above models of argument and explanation do not contain multiple perspectives.

Explanation and argumentation capabilities (Moulin et al, 2002) for more persuasive agents have already considered some aspects of user modeling. We have improved on this integration by also including the difference in the DL knowledge bases of agents. Fan and Toni (2015) have proposed a new argumentation semantic—related admissibility—designed to explain arguments. Fan and Toni have defined explanations as semantics, whereas we view explanation as a structured reason that is distinctive of arguments. The informal approach of Wright (2002) has been developed in this paper into a computational model of both argument and explanation.

Given different types of explanatory patterns in the social sciences (Miller, 2018), we limited our study to causal explanations. A broader investigation would include various types of explanations, such as *constructive explanations*, explaining events by accounting knowledge structures such as scripts and plans (Cashmore et al, 2019), or *contrastive explanations*, explaining surprising events by showing the deviation from expectation based on the available knowledge structures. Agents may convey even deceptive or rebellious explanations (Person and Person, 2019) like explanation with lying, explanation that holds information, explanation that is only a half-truth, cynical explanation, explanation with disobedience, or protest-based explanation. A plethora of explanations is now developing for explaining the black box models of deep learning. In this line, robust explanations aim to identify what is the smallest change to an instance to change decision (Shih, 2019). Minimum-cardinality explanation (Shih, 2019) identifies a minimal subset of the positive test results that is sufficient for the current decision. This broader spectrum of explanation requires to extend our ArgExp ontology. Such an extended ontology would support DL-based classification of explanations by reasoning on partial knowledge revealed at each step of the dialog.

Agents with Subjective Views.

Two individuals listening to the same debate may disagree regarding the winner of the dispute. Even when they hear the same arguments and corresponding attack relations, the agents can label differently the conveyed arguments. This may be due to the fact that the situation is approached from different perspectives that reflect the capabilities and experiences of each agent, because agents care about different criteria when determining the status of a conclusion (van der Weide et al, 2011).

An important issue in multi-agent systems is that of adaptability to other parties. While machine learning has been used to build the model of the opponent (Ledezma et al, 2009), in our approach the world of the opponent is inferred based on the speech acts used by this agent.

Cognitive maps follow the "personal construct theory" (Chaib-draa, 2002) providing a basis for the representation of individual multiple perspectives. The explanations in our model correspond to causal maps (Chaib-draa, 2002). From the perspective of modeling agent interactions, we consider that the model is more realistic when arguments are included.

Processing Natural Language Arguments.

Identifying structured arguments in natural language is the task of *argumentation mining*. Since 2014, argumentation mining has constantly attracted more researchers, as presented by Lawrence and Reed in their recent review (Lawrence and Reed, 2019). Differently, the complementary domain of *explanation mining* has not developed yet. However, explanation mining could rise as a subfield under the umbrella of explainable artificial intelligence (XAI) (Gunning, 2017).

Aware of the difficulty of argumentation mining (Debowska et al, 2009), we are engaged in the undertaking of building a bridge between natural dialog and its formal representation by using description logic. Differently from the argumentation schemes (Reed and Walton, 2007) or natural language processing (Wyner et al, 2012), our bridge intermingles two types of bricks: arguments and explanations. The proposed solution exploits human agent annotations to structure natural dialog according to the *ArgExp* common vocabulary.

From the dialog annotation perspective, the approach of the Twente Argumentation Schema (TAS) (Verbree et al, 2006) is similar to ours. In both cases, the developed tools allow users to annotate dialog based on a pre-defined vocabulary. TAS focuses on how statements involved in decision-making are related to each other, that is, on the structure of dialog. More narrowly, our goal was just to distinguish between argument and explanation. This narrow goal allowed us to define formally the *ArgExp* ontology. Differently from TAS, we assumed that the parties in dialog dynamically annotate their running conversation, and these annotations are used when deciding for the next move. By dynamically constructing their cognitive maps of the dialog, agents can react to misunderstanding as they occur in the conversation.

A relevant direction of modeling natural language arguments exploits argumentation schemes (Reed and Walton, 2007). A common element with our work is that each premise of a scheme has a specific type (Reed and Walton, 2007). For instance, *scheme from witness testimony* is supported by a premise of type *testimony*, *scheme from perception* by a premise of type *percept*, while *scheme from memory* is based on a premise of type *recollection*. In our case, one may have different types of evidence, like *direct evidence, circumstantial evidence, statistical evidence*, with the difference that they represent concepts in an ontology. This means that the reasoning tasks of DL can be exploited in our framework. A second difference is that we pay equal attention to explanations, which we consider important in natural dialogs. As argumentation schemes encapsulate patterns of human reasoning, their role of bridging the gap between low level formal models and natural dialog is essential. In this line, our approach is a starting point for formalizing *explanation schemes* similarly to the more investigated *argumentation schemes*.

In the explanatory argumentation framework in (Seselja and Strasser, 2013), Seselja and Strasser show how to apply abstract argumentation in scientific debates. We have been concerned here in mixing argument and explanation using DL knowledge so that human agents would be able to easily follow such a process. Therefore, our explanation was directed towards explaining on the knowledge level of the explainee, and not on explaining the workings of the abstract argumentation mechanism.

Two recent instruments aiming at filling the gap between natural language and our model of arguments and explanations are Targer (Chernodub et al, 2019) and Fred (Gangemi et al, 2017). Targer applies convolutional neural networks on labeled argument datasets in order to tag premises and conclusion in an argument. The distinction between argument and explanation is not considered, as the training datasets do not include explanations. A better mining model for our ArgExp ontology can be provided by the tool developed by Gangemi et al (2017). They have introduced the Fred tool aiming to automatically translate natural language into DL (Gangemi et al, 2017). Although Fred aims at general language, it might be a step towards a more specific instrument able to translate arguments and explanations into DL.

Argumentation and Description Logic.

A review of argumentation for the social Semantic Web has identified 14 Semantic Web models of argumentation (Schneider et al, 2013). These models are compared based on nine argumentation-related concepts: statement, issue, position, argument, causal, similarity, generic, supporting, and challenging. Three models include the notion of causal relation, which is in line with our notion of explanation. These are Semantic Annotation Vocabulary, ScholOnto and LKIF (Schneider et al, 2013). None of these three models has explicitly defined the concept of argument (Schneider et al, 2013). In our approach, we bind arguments and explanations under the same umbrella of the *ArgExp* ontology.

The most referred model for DL-based arguments remains the Argument Interchange Format (AIF) ontology, which represents the foundation of the World Wide Argumentative Web. This argumentative web was envisaged as a large-scale interconnection of structured arguments posted by human agents on the Web (Rahwan, 2008). Relevant extensions of AIF deal with representation of argumentation schemes (Rahwan et al, 2007) and of dialogical argumentation (Modgil and McGinnis, 2007; Reed et al, 2008). From this perspective, our work can be seen as an extension of AIF with explanations, with the focus on distinguishing between argument and explanation. Our solution addresses the distinction at the level of concepts in an ontology, but also at the level of speech acts used to convey arguments or explanations.

6 Conclusions

This paper formalizes a precise distinction between argument and explanation in a dialogue, and models it in Description Logics. Given the ubiquity of arguments and explanations in natural dialog, our contributions are: (i) providing guidelines to determine whether something in a dialog is an argument or an explanation; (ii) modeling explanations and arguments in description logic within the same *ArgExp* ontology. (iii) modeling subjective perspective of agents on arguments and explanation. By exploiting the reasoning tasks of the DL, the system we implemented is able to automatically classify arguments and explanations, based on the partial information disclosed during dialog. The main benefit is that agents identify more quickly agreements and disagreements during dialogs.

We claim that our model may have applicability in the following areas. (i) In *legal discourses*, distinguishing between argument and explanation provides insights on the pleading games (Gordon, 1993). Our model allows the integration of legal ontologies for handling refined types of legal evidence. (ii) In *press articles*, our formalization is a step toward semi-automatic identification of the structure, as informally suggested in (Mayes, 2010). (iii) In *learning*, the use of such a system would be to structure argumentation and explanation for understanding scientific notions (de Vries et al, 2002) using computer-mediated dialogs tools enriched with semantic annotation. (iv) In the *standards for dialog annotation*, by exploiting the semantics of RDF or OWL instead of XML used for the ISO 24617-2 dialog annotation standard (Bunt et al, 2017), it would be easier to build applications that conform to the standard.

Our computational model may be extended in several directions. First, our approach can be seen as a starting point for defining an ontology of explanations, complementary to – and completing in our view – the AIF argumentation ontology. Of course, explanations should not be limited to our causal model here, but also to include other types of explanations (e.g. counterfactual). Second, one can investigate how does the model fit to dialogs with more than two agents, like open discussions. What about the situation in which a mediator exists, aware of the objective world w_O? It would be interesting to compare how disagreement decreases (Booth et al, 2012) as the dialog evolves: (i) with and without a mediator and (ii) with and without explanation capabilities. Third, it would be interesting to anlayse how the agents are shifting between cooperative dialogues (i.e. explanatory dialogues in our case) and competitive dialogues (i.e. persuasive or argumentative dialogues in our approach).

Acknowledgments

We are grateful for the useful comments from anonymous reviewers and participants at the Poznań Reasoning Week 2018. Adrian Groza is supported by the ExNanoMat-21PFE grant.

References

Amgoud L, Serrurier M (2008) Agents that argue and explain classifications. Autonomous Agents and Multi-Agent Systems 16(2):187–209

Arioua A, Buche P, Croitoru M (2017) Explanatory dialogues with argumentative faculties over inconsistent knowledge bases. Expert Systems with Applications 80:244–262

Bex F, Prakken H (2008) Investigating stories in a formal dialogue game. In: Conference on Computational Models of Argument, IOS Press, pp 73–84

Bex F, Walton D (2016) Combining explanation and argumentation in dialogue. Argument & Computation 7(1):55–68

Bex F, Van Koppen P, Prakken H, Verheij B (2010) A hybrid formal theory of arguments, stories and criminal evidence. Artificial Intelligence and Law 18(2):123–152

Booth R, Caminada M, Podlaszewski M, Rahwan I (2012) Quantifying disagreement in argument-based reasoning. In: AAMAS, pp 493–500

Brem S, Rips L (2000) Explanation and evidence in informal argument. Cognitive Science 24:573–604

Budzynska K, Reed C (2011) Speech acts of argumentation: Inference anchors and peripheral cues in dialogue. In: Computational Models of Natural Argument

Bunt H, Petukhova V, Traum D, Alexandersson J (2017) Dialogue act annotation with the iso 24617-2 standard. In: Multimodal interaction with W3C standards, Springer, pp 109–135

Cashmore M, Collins A, Krarup B, Krivic S, Magazzeni D, Smith D (2019) Towards explainable AI planning as a service. arXiv preprint arXiv:190805059

Chaib-draa B (2002) Causal maps: Theory, implementation, and practical applications in multiagent environments. IEEE Transactions on Knowledge and Data Enginnering 14(6):1201–1217

Chernodub A, Oliynyk O, Heidenreich P, Bondarenko A, Hagen M, Biemann C, Panchenko A (2019) Targer: Neural argument mining at your fingertips. In: Proceedings of the 57th Annual Meeting of the Association for Computational Linguistics: System Demonstrations, pp 195–200

Debowska K, Lozinski P, Reed C (2009) Building bridges between everyday argument and formal representations of reasoning. Studies in Logic, Grammar and Rhetoric 16:95–135

Fan X, Toni F (2015) On computing explanations in argumentation. In: AAAI, pp 1496–1502

Gangemi A, Presutti V, Reforgiato Recupero D, Nuzzolese AG, Draicchio F, Mongiovì M (2017) Semantic web machine reading with Fred. Semantic Web 8(6):873–893

Gordon TF (1993) The pleadings game. Artificial Intelligence and Law 2(4):239–292

Gordon TF, Prakken H, Walton D (2007) The Carneades model of argument and burden of proof. Artificial Intelligence 171(10-15):875–896

Groza A (2018) Distinguishing argument and explanation with description logic. In: Poznan Reasoning Week, Games and Reasoning, Logic & Cognition, Refutation Symposium, 11–15 September 2018, Poznan, Poland, pp 31–32

Guid M, Možina M, Pavlič M, Turšič K (2019) Learning by arguing in argument-based machine learning framework. In: International Conference on Intelligent Tutoring Systems, Springer, pp 112–122

Gunning D (2017) Explainable artificial intelligence (XAI). Defense Advanced Research Projects Agency (DARPA), nd Web 2

Hempel CG, Oppenheim P (1948) Studies in the logic of explanation. Philosophy of Science 15(2):135–175

Hilton DJ (1990) Conversational processes and causal explanation. Psychological Bulletin 107(1):65

Janvid M (2012) Knowledge versus understanding: The cost of avoiding Gettier. Acta Analytica 27:183–197

Lawrence J, Reed C (2019) Argument mining: A survey. Computational Linguistics (Just Accepted):1–55

Ledezma A, Aler R, Sanchís A, Borrajo D (2009) OMBO: An opponent modeling approach. AI Communications 22(1):21–35

Letia IA, Groza A (2012) Interleaved argumentation and explanation in dialog. In: Computational Models of Natural Argument, pp 44–52

Lumer C (2005) The epistemological theory of argument: How and why? Informal Logic 25:213–242

Mayes GR (2000) Resisting explanation. Argumentation 14:361–380

Mayes GR (2010) Argument explanation complementarity and the structure of informal reasoning. Informal Logic 30(1):92–111

McBurney P, Parsons S (2001) Representing epistemic uncertainty by means of dialectical argumentation. Annals of Mathematics and Artificial Intelligence 32(1-4):125–169

Mercier H, Sperber D (2011) Why do humans reason? Arguments for an argumentative theory. Behavioral and Brain Sciences (34):57–111

Miller T (2018) Explanation in artificial intelligence: Insights from the social sciences. Artificial Intelligence

Modgil S, McGinnis J (2007) Towards characterising argumentation based dialogue in the argument interchange format. In: ArgMAS, pp 80–93

Moulin B, Irandoust H, Belanger M, Desbordes G (2002) Explanation and argumentation capabilities: Towards the creation of more persuasive agents. Artificial Intelligence Review 17:169–222

Paglieri F, Castelfranchi C (2010) Why argue? Towards a cost-benefit analysis of argumentation. Argument and Computation 1:71–91

Pearl J, Mackenzie D (2018) The book of why: the new science of cause and effect. Basic Books

Person A, Person B (2019) When agents talk back: Rebellious explanations

Poole D (1989) A methodology for using a default and abductive reasoning system. Tech. rep., Vancouver, BC, Canada, Canada

Rahwan I (2008) Mass argumentation and the Semantic Web. Jornal of Web Semantics 6(1):29–37

Rahwan I, Zablith F, Reed C (2007) Laying the foundations for a World Wide Argument Web. Artificial Intelligence 171(10-15):897–921

Reed C, Walton D (2007) Argumentation schemes in dialogue. In: Hansen HV (ed) Dissensus and the Search for Common Ground, pp 1–11

Reed C, Wells S, Devereux J, Rowe G (2008) AIF+: Dialogue in the argument interchange format. In: COMMA, pp 311–323

Schneider J, Groza T, Passant A (2013) A review of argumentation for the social semantic web. Semantic Web: Interoperability, Usability, Applicability In press

Seselja D, Strasser C (2013) Abstract argumentation and explanation applied to scientific debates. Synthese In press

Shanahan M (1989) Prediction is deduction but explanation is abduction. In: Proceedings of the 11th International Joint Conference on Artificial Intelligence, Morgan Kaufmann, pp 1055–1060

Shih B (2019) Explaining classifiers. PhD thesis, UCLA

Tempich C, Pinto H, Sure Y, Staab S (2005) An argumentation ontology for DIstributed, Loosely-controlled and evolvInG Engineering processes of oNTologies (DILIGENT). In: ESWC, pp 241–256

Verbree D, Rienks R, Heylen D (2006) First steps towards the automatic construction of argument-diagrams from real discussions. In: COMMA, pp 183–194

de Vries E, Lund K, Baker M (2002) Computer-mediated epistemic dialogue: Explanation and argumentation as vehicles for understanding scientific notions. Journal of the Learning Sciences 11(1):63–103

Walton D (2011) A dialogue system specification for explanation. Synthese 182(3):349–374

Walton D, Godden DM (2007) Redefining knowledge in a way suitable for argumentation theory. In: Dissensus and the Search for Common Ground: Proceedings of OSSA, pp 1–13

van der Weide T, Dignum F, Meyer JJC, Prakken H, Vreeswijk G (2011) Multi-criteria argument selection in persuasion dialogues. In: AAMAS, pp 921–928

Wright L (2002) Reasoning and explaining. Argumentation 16:33–46

Wyner A, Schneider J, Atkinson K, Bench-Capon TJM (2012) Semi-automated argumentative analysis of online product reviews. In: COMMA, pp 43–50

Xu J, Yao L, Li L, Ji M, Tang G (2020) Argumentation based reinforcement learning for meta-knowledge extraction. Information Sciences 506:258–272

Zeng Z, Fan X, Miao C, Leung C, Jih CJ, Soon OY (2018) Context-based and explainable decision making with argumentation. In: Proceedings of the 17th International Conference on Autonomous Agents and MultiAgent Systems, International Foundation for Autonomous Agents and Multiagent Systems, pp 1114–1122

A Formal Ontology for Conception Representation in Terminological Systems

Farshad Badie

Abstract I have supposed that we need a formal system to represent and explicate humans' conceptions of the world. According to this research, such a formal system is representable based on a Conception Language ($\mathcal{CL}$) that is a terminological knowledge representation formalism. In this research, I will offer a formal ontology for conception representation in terminological systems. Such a $\mathcal{CL}$-based ontology will specify the conceptualisation of humans' conceptions as well as of the effects of their conceptions on the world.

Key words: concept, conception, conception language, Description Logics, ontologies, terminological systems

1 Introduction

Our conceptions of the world are formed primarily by engaging with our environment as well as based on our interactions with other agents (e.g., human beings, animals, artificial agents). Conceptions are in an important sense expressed in the form of our descriptions, statements, specifications, explanations, justifications, questions, answers and requests. Conceptions have presuppositions in their favours. However, they are open to being revised and updated. This means that a conception (of the world) is corrigible and is capable of being annulled. For example, Bob may have a conception of 'autumn' that can be expressed in his relevant descriptions and explanations. Correspondingly, his conception of autumn becomes manifested in any of his relevant questions, answers, justifications, etc. In addition, he can/may update his conception of 'autumn' with regard to his research, experiments and conversational exchanges.

Farshad Badie
Center for Natural and Formal Languages
Aalborg University, Denmark
e-mail: badie@id.aau.dk

In the recent decades, knowledge representation in information and computer sciences has experienced significant improvements, see Brachman and Levesque (1985); Levesque and Brachman (1985); van Harmelen et al (2008). The underlying Description Logics (DLs) are now among the most widely used knowledge representation formalisms in terminological as well as semantics-based systems, see Baader (2017); Sikos (2017); Baader et al (2010); Rudolph (2011); Buchheit et al (1993). DLs have emerged from semantic networks Quillian (1968) and frame-based systems Minsky (1974). Note that most members of the family of description logics are decidable fragments of predicate logic (PL). Thus, It can be interpreted that DLs are PL-based terminological systems developed out of the attempt to represent (terminological) knowledge, with a formal semantics, in order to establish a common ground for human and machine interplays.

This research designs a Conception Language ($\mathcal{CL}$) that is a kind of description logic. $\mathcal{CL}$ is utilised as a formal system for representing and explicating humans' conceptions (of the world). The main focus of this paper is on conceptualising conceptions and, subsequently, on specifying the conceptualisation of conceptions in the framework of $\mathcal{CL}$. Correspondingly, the final and the most significant contribution of the research is to offer a formal (as well as a schematic) ontology for humans' conceptions within terminological systems.

2 What are Concepts?

There has always been a general problem concerning the sense and notion of the concept of 'concept', in linguistics, psychology, cognitive science, philosophy, and computer science. Over the years, the term 'concept' has not been used consistently, see Bartlett (1932); Allwood and Andersson (1976); Peacocke (1992); A. and E. Moss (2003); Margolis and Laurence (2007, 2010); Margolis (2015). In Kant's words, a concept is the "unity of the act of bringing various representations under one common representation", see Kant (1781/1964). Kant argued that "no concept is related to an object immediately, but only to some other representation[s] of it". In this sense, concepts offer a linkage between the mental representations of linguistic expressions and the other mental images (e.g., representations of the world, representations of inner experiences) that one has in his mind, see Badie (2017b). According to Allwood (1999); Allwood and Andersson (1976), we can discern three fundamental positions that apply to what a concept really is:

1. Concepts are abstract phenomena which exist outside and are independent of space and time. This position is called *conceptual realism* and is currently espoused by many mathematicians and logicians. According to conceptual realism, concepts are eternal and constant phenomena. Subsequently, concepts are assumed to have a real existence. Conceptual realism was first advocated by Plato.
2. Concepts are mental phenomena that are discovered and construed by human beings. This position is usually called *conceptualism*. Conceptualism is currently

the most common view among psychologists and linguists. Conceptualism was first advocated by Aristotle and the medieval semanticist Abelard.
3. Concepts do not really exist. The only things that exist are (i) linguistic expressions and (ii) things in the world that linguistic expressions can represent. In this view, our direct experience of the world is often identified with the world itself. This third position is usually called *nominalism*. According to nominalism, concepts are reduced to a name (or rather a linguistic expression) or even a symbol. Currently, nominalistic views are found among linguists, philosophers, logicians and psychologists who are influenced by so-called 'behaviourism'. Nominalism originally emerged in antiquity in opposition to the Platonic conceptual realism. One of nominalism's most well-known spokesmen, who lived much later, is the medieval philosopher Occam.

This research is relied on my assumption that 'we—in order to conceptualise and assess a concept from the perspective of formal logic—need to consider any concept as (a) a conceptual entity and (b) a logical-assessable phenomenon (e.g., mathematical phenomenon)'. Subsequently, concepts can be applied to different contexts, e.g., knowledge representation systems, knowledge acquisition systems, smart learning systems.

3 What are Conceptions?

It shall be taken into consideration that this research is relied on the following psychological suppositions:

1. Concepts are the final products of one's conceptualisation of the world.
2. Concepts are originally mental phenomena that are conceived in (and produced by) human beings' consciousness.

It is interpretable that the afore-mentioned psychological suppositions are in line with conceptualism. However, this paper does not deal with the ontology of concepts, concept production processes, and concepts' existence(s) in the mind. Instead, the paper focuses on the formal-logical analysis of concepts within possible descriptions of the world in terminological systems. In more proper words, this research is relied on '*Nominal Conceptualism*'. In my opinion, concepts are the primary and fundamental units of humans' (terminological) knowledge and are the basic materials of humans' constructed meanings. Concepts can be identified with the contents in certain means of expression, e.g., linguistic expression, formal expression. I thus believe that human beings—for expressing their own world descriptions based on their mental 'concepts'—deal with their linguistic expressions that are said/written based on, e.g., letters, symbols. In fact, humans become concerned with the production of their 'conceptions' of the world. Thereby, in my theoretical model, conceptions are the consequences of concepts (that are conceptual entities). Concepts are the basic materials of meanings. Subsequently, meanings are regarded as 'conceptual structures'.

In this research I am going to deal with conception representation in terminological systems. It is an underlying research hypothesis of the paper that the most fundamental building blocks of formal world descriptions are expressible based on conceptions (of concepts). For the logical assessment of conceptions, any conception (of a concept) and its interconnections with other conceptions can—by means of predication—be assigned into logical symbols, see Badie (2017a). For example, for the logical-terminological assessment of one's conception of the concept 'tree', the conceptual entity *Tree* and its logical-terminological interrelationships with any other conceptual entity can be predicated.

4 $\mathcal{CL}$: A Formal Language for Expressing Conceptions

We need a formal system to explicate humans' conceptions of the world. Such a formal system can be represented based on *Conception Language* ($\mathcal{CL}$) that is a description logic and a terminological knowledge representation formalism. $\mathcal{CL}$ (as a decidable fragment of predicate logic) represents knowledge in terms of (1) conceptions, (2) singulars, and (3) effects.

1. A *conception* (of a concept) corresponds to a [distinct] conceptual entity (that might be also regarded as a class of other conceptual entities). Conceptions and their interrelationships (that are connected to each other in the form of hierarchical structures) create terminologies. From the perspective of predicate logic, one's conception of a concept can be interpreted to be equivalent to a unary predicate. Atomic conceptions (e.g., conception of *Tree*, conception of *Bird*, conception of *Food*, conception of *Information*) are the first group of atomic symbols in $\mathcal{CL}$.
2. *Singulars* are the instances of the conceptions (of concepts). Thereby it is realisable that conceptions can describe singulars. For example, the singular bob (who is conceptualised to be a student) is describable as an instance of the conception of the concept *Student*. Actually the singular bob is covered by the conception of *Student*. Assessed by predicate logic, singulars are equivalent to constant symbols and can be regarded as 0-ary predicates. Singulars are the second group of atomic symbols in $\mathcal{CL}$.
3. An *effect* (of a conception of a concept like *C*) expresses a relationship between a singular (that is conceptualised to be the instance of *C*) and other singulars (that are conceptualised to be the instances of either *C* or any other concept). In addition, an effect (of a conception of *C*) can assign a property to a singular (that is covered by *C*). Correspondingly, it is interpreted that effects assign attributes to singulars. It shall be concluded that effects are the relations defined with some valence greater than or equal to 2. Effects are logically equivalent to *n*-ary predicates. Atomic effects (e.g., *hasSon, isTeaching, isMoving, hasProduct, isBasedOn*) are the third group of atomic symbols in $\mathcal{CL}$.

Here are the examples of world descriptions based on the effects of the valences 0, 1 and 2. Obviously, there can be relations of the greater valences as well.

A Formal Ontology for Conception Representation in Terminological Systems   141

- Any of the singulars ann, google, peach, blue, and october are related to itself by means of the effect of valence 0.
- The terms 'Fred is a student' (formally speaking: *Student*(fred)), 'Anna is a mother' (or: *Mother*(anna)), 'Google is a company' (or: *Company*(google)), and 'Blue is a Colour' (or: *Colour*(blue)) are structured based on the effects of valence 1. It can be contemplated that any effect of valence 1 stands for the term 'is a'. In fact, the effects of valence 1 express the concept of 'being'. According to *Student*(fred), the singular fred has been conceptualised and described based on the conception of *Student*.
- The terms 'Tom is married to Juliana' (formally: *marriedTo*(tom, juliana)), '10 is greater than 3' (or: *greaterThan*(10, 3)), and 'Bob is the father of Alice' (or: *hasFather*(alice, bob)) are structured based on the effects of valence 2. According to *hasFather*(alice, bob), the singulars alice and bob (that are the instances of the conceptions of two concepts) are conceptualised to be related to each other by means of the effect *hasFather*. In other words, the singular alice has—by means of the effect *hasFather*—effected the singular bob. Equivalently, the singular bob has—by means of the effect *hasFather*—been effected by the singular alice.

4.1 Syntax

4.1.1 Non-Logical Symbols

Atomic symbols are the most fundamental descriptions from which human beings, inductively, build more-specified as well as complex world descriptions. Thereby it shall be concluded that our formalism is strongly dependent on the collection(s) of:

1. Atomic conceptions (that are the conceptions of atomic concepts),
2. Singulars (that are the instances of the conceptions of atomic concepts), and
3. Atomic effects (that are the interrelationships between the instances of the conceptions of atomic concepts).

Regarding $N_\mathcal{A}, N_\mathcal{E}$, and N_S as the sets of atomic conceptions, atomic effects, and singulars, respectively, the triple $\langle N_\mathcal{A}, N_\mathcal{E}, N_S \rangle$ denotes a *signature*. Note that the ingredients of signatures do not independently have any logical consequence in a $\mathcal{CL}$-based world description.

4.1.2 Logical Symbols

The set of main logical symbols in $\mathcal{CL}$ is {conjunction ($\sqcap$), disjunction ($\sqcup$), negation ($\neg$), implication ($\rightarrow$), existential quantification ($\exists$), universal quantification ($\forall$)}. In addition, the formalism, logically, supports {tautology ($\top$), contradiction ($\bot$)}. More specifically, the logical symbols $\top$ and $\bot$ express the concepts of 'tautology' (the truth of a conception of a concept based on all possible interpretations) and 'contradiction' (the falsity of a conception of a concept based on all possible interpretations), respectively.

$\mathcal{CL}$ is a PL-based terminological system. However, it is assumed that $\mathcal{CL}$ is primarily a syntactic variant of the modal logic $\mathbf{K}$[1]. Let me be more specific. According to one's conceptions of the concepts C and D, the logical descriptions $\neg C$ (conception negation) and $C \to D$ (conception implication) can be regarded as the translations of the $\mathbf{K}$'s logical symbols '$\sim$' (negation) and '$\Rightarrow$' (implication). Also, the $\mathcal{CL}$ descriptions $C \sqcap D$ (conception conjunction) and $C \sqcup D$ (conception disjunction) are definable from $\mathcal{CL}$ descriptions $\neg C, \neg D, C \to D$, and $D \to C$ in the same way as in propositional logic and, in fact, as in $\mathbf{K}$ has been defined. Actually $\mathbf{K}$ has defined its logical symbols '&' (conjunction) and '∨' (disjunction) from the logical symbols '$\sim$' and '$\Rightarrow$'.

4.2 Semantics

The formal semantics in $\mathcal{CL}$ is analysable based on the interpretations of the ingredients of signatures (that are non-logical symbols) in different descriptions. In fact, we—in order to deal with the formal semantics of conceptions (as well as their effects)—need to utilise terminological interpretations. A terminological interpretation (like $\mathcal{I}$) is offered based on the following items:

1. *The non-empty set $\mathcal{D}$.* This set is the domain of the interpretation of a conception. It consists of any possible singular that can occur in any relevant description based on that conception.
2. *The interpretation function '$.^{\mathcal{I}}$'.* This function transforms any singular symbol (like a) into its interpreted form (like $a^{\mathcal{I}}$), such that $a^{\mathcal{I}} \in \mathcal{D}^{\mathcal{I}}$. In addition, the interpretation function transforms any atomic conception $\mathcal{A}$ into the set $\mathcal{A}^{\mathcal{I}}$, such that $\mathcal{A}^{\mathcal{I}} \in \mathcal{D}^{\mathcal{I}}$. It also transforms any atomic effect $\mathcal{E}$ into the n-ary relation $\mathcal{E}^{\mathcal{I}}$, such that $\mathcal{E}^{\mathcal{I}} \subseteq \mathcal{D}^{\mathcal{I}} \times \mathcal{D}^{\mathcal{I}} \times \cdots \times \mathcal{D}^{\mathcal{I}}$.

Table 1 represents the syntax and semantics of fundamental descriptions in $\mathcal{CL}$. In this table, C and D are two conceptions (of two concepts), $\mathcal{A}$ is an atomic conception, $\mathcal{E}$ is an atomic effect, and a and b are two singulars.

5 Terminological and Assertional Axiomatisation in $\mathcal{CL}$

In our formalism, conceptions and the interrelationships between singulars (by means of effects) are—in the form of hierarchical structures—employed in order to create terminologies. It has been taken into account that our conceptions of the world are terminologically describable in the forms of (1) conception subsumption (or equivalently,

---

[1] $\mathbf{K}$ was named after Saul Aaron Kripke, who is an American logician and philosopher. Kripke is well-known for his valuable works on the semantics of modal logic, see Patrick Blackburn (2006).

Table 1: $\mathcal{CL}$ Syntax and Semantics

Syntax	Semantics
$\mathcal{A}$	$\mathcal{A}^{\mathcal{I}} \subseteq \mathcal{D}^{\mathcal{I}}$
$\mathcal{E}$	$\mathcal{E}^{\mathcal{I}} \subseteq \mathcal{D}^{\mathcal{I}} \times \mathcal{D}^{\mathcal{I}} \times \cdots \times \mathcal{D}^{\mathcal{I}}$
$\top$	$\mathcal{D}^{\mathcal{I}}$
$\bot$	$\varnothing$
$C \sqcap D$	$(C \sqcap D)^{\mathcal{I}} = C^{\mathcal{I}} \wedge D^{\mathcal{I}}$
$C \sqcup D$	$(C \sqcup D)^{\mathcal{I}} = C^{\mathcal{I}} \vee D^{\mathcal{I}}$
$\neg C$	$(\neg C)^{\mathcal{I}} = \mathcal{D}^{\mathcal{I}} \setminus C^{\mathcal{I}}$
$C \rightarrow D$	$D^{\mathcal{I}} \vee (\neg C)^{\mathcal{I}}$
$\exists \mathcal{E}.C$	$\{a \mid \exists b.(a,b) \in \mathcal{E}^{\mathcal{I}} \wedge b \in C^{\mathcal{I}}\}$
$\forall \mathcal{E}.C$	$\{a \mid \forall b.(a,b) \in \mathcal{E}^{\mathcal{I}} \rightarrow b \in C^{\mathcal{I}}\}$

conception of concept subsumption), (2) conception equivalency (or equivalently, conception of concept equivalency), (3) effect subsumption (or equivalently, conception of relation subsumption), and (4) effect equivalency (or equivalently, conception of relation equivalency). However, (3) and (4) are highly dependent on (1) and (2), respectively. Furthermore, in our formalism, humans' conceptions of the world are assertionally describable in the forms of (5) conception assertions (or equivalently, conception of concept assertions) and (6) effect assertions (or equivalently, conception of relation assertions). In this research, a terminological interpretation is called a *model* for a $\mathcal{CL}$-based description if it can satisfy all the terminological and assertional axioms based on which that description has been expressed. Let me be more specific on terminological and assertional axioms:

1. *Conception subsumption.* According to the conception subsumption $C \sqsubseteq D$, it is conceptualised that the conception C (that is, in fact, the conception of the concept C^2) is subsumed under the conception D. Semantically, $C^{\mathcal{I}} \subseteq D^{\mathcal{I}}$. In more proper words, it is interpreted that C is the sub-conception of D. For example, regarding John's conception, 'leafs are plants' (I shall emphasise that I am not dealing with the truth/falsity of John's conception, but only with the logical-terminological structure of his conception). Formally speaking, $Leaf \sqsubseteq Plant$. According to this fundamental logical-terminological description, John's conception of *Leaf* is subsumed under his conception of *Plant*. Semantically, $Leaf^{\mathcal{I}} \subseteq Plant^{\mathcal{I}}$.

2. *Conception equivalency.* According to the conception equivalency $C \equiv D$, it is conceptualised that the conception C is equivalent to the conception D. Semantically, $C^{\mathcal{I}} = D^{\mathcal{I}}$. Hence it is interpreted that the conception C is equal to the conception D. For example, regarding Mary's conception, the concepts 'husband' and 'spouse' are equivalent. In fact, her conception of *Husband* is equivalent to her conception of *Spouse*. Formally speaking, $Husband \equiv Spouse$. Semantically, $Husband^{\mathcal{I}} = Spouse^{\mathcal{I}}$.

3. *Effect subsumption.* According to the effect subsumption $E \sqsubseteq F$, it is conceptualised that the effect E is subsumed under the effect F. Semantically, $E^{\mathcal{I}} \subseteq F^{\mathcal{I}}$.

---

[2] In this research, 'the conception C' is equivalent to 'the conception of the concept C'.

Therefore, it is interpreted that E is the sub-effect of F. For example, James has conceptualised that 'learning is memorising'. For instance, he may conceptualise that 'learning history by Tommy' is 'memorising history by Tommy'. Formally speaking, *isLearning*(tommy, history) $\sqsubseteq$ *isMemorising*(tommy, history). According to such a fundamental logical-terminological description based on this conception, 'learning history by Tommy' is subsumed under 'memorising history by Tommy'. This means that *isLearning*$^\mathcal{I}$(tommy$^\mathcal{I}$, history$^\mathcal{I}$) $\subseteq$ *isMemorising*$^\mathcal{I}$(tommy$^\mathcal{I}$, history$^\mathcal{I}$). However, as I mentioned above, effect subsumptions are dependent on conception subsumptions. Hence, it should be taken into account that 'the conception of the effect of the conception of *Learning*' is subsumed under 'the conception of the effect of the conception of *Memorisation*'. Consequently, the effect subsumption *isLearning*(tommy, history) $\sqsubseteq$ *isMemorising*(tommy, history) is dependent on the conception subsumption *Learning* $\sqsubseteq$ *Memorisation*. Utilising the terminological interpretation $\mathcal{I}$, we formally have:

$$(Learning^\mathcal{I} \subseteq Memorisation^\mathcal{I})$$
$$\rightarrow$$
$$(isLearning^\mathcal{I}(\text{tommy}^\mathcal{I}, \text{history}^\mathcal{I}) \subseteq isMemorising^\mathcal{I}(\text{tommy}^\mathcal{I}, \text{history}^\mathcal{I})).$$

4. *Effect equivalency.* According to the effect equivalency $E \equiv F$, it is conceptualised that the effect E is equivalent to the effect F. Semantically, $E^\mathcal{I} = F^\mathcal{I}$. Therefore, it is interpreted that E is equal to F. For example, Ann may conceptualise that 'observing something is equivalent to seeing that thing'. For instance, she may conceptualise that 'observing the apple by Mary' is equivalent to 'seeing the apple by Mary'. Formally speaking, *isObserving*(mary, apple) $\equiv$ *isSeeing*(mary, apple). According to such a fundamental logical-terminological description based on Ann's conception, there is an effect equivalency between 'observing the apply by Mary' and 'seeing the apply by Mary'. Semantically, *isObserving*$^\mathcal{I}$(mary$^\mathcal{I}$, apple$^\mathcal{I}$) = *isSeeing*$^\mathcal{I}$(mary$^\mathcal{I}$, apple$^\mathcal{I}$). However, as I mentioned above, effect equivalencies are dependent on conception equivalencies. Correspondingly, 'the conception of the effect of the conception of *Observation*' is equivalent to 'the conception of the effect of the conception of *Sight*'. Subsequently, the effect equivalency *isObserving*(mary, apple) $\equiv$ *isSeeing*(mary, apple) is dependent on the conception equivalency *Observation* $\equiv$ *Sight*. Utilising the terminological interpretation $\mathcal{I}$, we formally have:

$$(Observation^\mathcal{I} = Sight^\mathcal{I})$$
$$\rightarrow$$
$$(isObserving^\mathcal{I}(\text{mary}^\mathcal{I}, \text{apple}^\mathcal{I}) = isSeeing^\mathcal{I}(\text{mary}^\mathcal{I}, \text{apple}^\mathcal{I})).$$

5. *Conception assertion.* According to the conception assertion $C(a)$, it is conceptualised that the singular a is described by the conception C. Semantically, $a^\mathcal{I} \in C^\mathcal{I}$. Actually it is interpreted that the singular a (which has been conceptualised and interpreted by a human being) belongs to the interpreted C. For example, regarding Mary's conception, *Husband* $\equiv$ *Spouse*. Therefore, her husband, Brian, is also her spouse. Semantically, *Husband*$^\mathcal{I}$ = *Spouse*$^\mathcal{I}$. Consequently,

$$(\text{brian}^\mathcal{I} \in Husband^\mathcal{I}) \rightarrow (\text{brian}^\mathcal{I} \in Spouse^\mathcal{I}).$$

Similarly,

$$(\text{brian}^{\mathcal{I}} \in \textit{Spouse}^{\mathcal{I}}) \rightarrow (\text{brian}^{\mathcal{I}} \in \textit{Husband}^{\mathcal{I}}).$$

6. *Effect assertion.* According to the effect assertion $E(a_1, a_2, ..., a_n)$, it is conceptualised that all the singulars $a_1, a_2, ..., a_n$ are related together by means of the effect E. Semantically, $(a_1^{\mathcal{I}}, a_2^{\mathcal{I}}, ..., a_n^{\mathcal{I}}) \in E^{\mathcal{I}}$. In fact, the collection of the interpretations of n singulars belongs to the interpretation of E. For example, regarding Barbara's conception—that *Leaf* $\sqsubseteq$ *Plant*—a leaf of her rose flower is an instance of her conception of the concept *Plant*. Formally speaking, *Plant*(roseLeaf). In fact, her conception of *Plant* can describe her conception of her rose's leaf. Semantically,

$$\text{roseLeaf}^{\mathcal{I}} \in \textit{Leaf}^{\mathcal{I}}$$
$$\land$$
$$\textit{Leaf}^{\mathcal{I}} \subseteq \textit{Plant}^{\mathcal{I}}$$

Therefore:

$$\text{roseLeaf}^{\mathcal{I}} \in \textit{Plant}^{\mathcal{I}}.$$

6 Conception Language: Main Objectives

$\mathcal{CL}$ is a conception representation formalism. It provides a formal background for expressing conceptions of the world. The following items are the most important objectives of a conception representation formalism.

I. Dealing with Conception/Effect Construction. Conception construction focuses on the production as well as the development of new conceptions based on available conceptions of the world. Conception constructions are usually produced in the following forms:

1. *Conception implication* deals with the implication of (the existence of) a conception from one's available conceptions of the world. For example, one can construct his conception of 'mentor' based on his conception of 'teacher'. Formally speaking, *Teacher* → *Mentor*.
2. Relying on conception implications, *effect implications* are analysable. For example, considering Mary's conception, Bob (who has digested an apple), has certainly eaten an apple. Formally speaking, *isDigesting*(bob, apple) → *isEating*(bob, apple).
3. *Conception opposition* focuses on the construction of the opposition of one's available conceptions of the world. For example, one can—based on his conception of 'tree' (formally: *Tree*)—produce his conception of 'not-tree' (formally: ¬*Tree*).
4. Relying on conception oppositions, *effect oppositions* are analysable. For example, it is possible to—based on the conception of *isLearning*(tom, history)—produce the conception of ¬*isLearning*(tom, history).

5. *Conception junction* focuses on the production of the junction of two (or more) available conceptions of the world. One can—based on his conceptions of 'circle', 'big' and 'blue'—produce his conception of 'big blue circle'. It formally can be described by *Circle* $\sqcap$ *Big* $\sqcap$ *Blue*.
6. Regarding conception junctions, *effect junctions* are analysable. For example, it is possible to—based on the conceptions of *isReading*(bob, philosophy) and *isSmiling*(bob, mary)—produce the conception of *isReading*(bob, philosophy) $\sqcap$ *isSmiling*(bob, mary).
7. *Conception unification* focuses on the production of the unification of two (or more) available conceptions of the world. One can—based on his conceptions of 'spring' and 'autumn'—select either Spring or Autumn as the season of his wedding ceremony. It is formally described by *Spring* $\sqcup$ *Autumn*.
8. Regarding conception unifications, *effect unifications* are analysable. Maria can—based on her conceptions of 'studying psychology' and 'playing chess'—decide if she chooses to study or play. Formally speaking, *isStudying*(maria, psychology) $\sqcup$ *isPlaying*(maria, chess).
9. *Conception quantification* that quantifies the available conceptions of the world. One can—based on his conception of his 'students'—select either 'some' or 'all' of his students (formally, $\exists$ *Student* or $\forall$ *Student*) for his new research project.
10. Conception quantifications and *effect quantifications* are strongly tied together. It means that the quantification of one's conceptions is dependent on the quantification of his conceptions' properties and, in fact, on the quantification of his conceptions' effects (on his other conceptions). For example, one (who has selected either 'some' or 'all' of his students for a new research project) has quantified his conception of the 'students' as well as of 'their properties' (e.g., their skills, their talents, their attributes).
11. *Conception qualification* that qualifies the available conceptions of the world. One can—based on his conception of 'flower'—make his conception of 'beautiful flower'. Note that conception qualification is strongly related to conception conditioning (see item VIII).
12. Conception qualifications and *effect qualifications* are strongly tied together. It means that the qualification of one's conceptions is dependent on the qualification of his conceptions' properties and, in fact, on the qualification of his conceptions' effects (on his other conceptions). For example, one (who has made his conception of 'beautiful flower') has qualified his conception of 'flower' as well as of 'its properties' (e.g., flowers' colours, flowers' shapes).

II. **Dealing with Conception/Effect Classification.** Conception classification classifies the available conceptions of the world. According to Bob's conceptualisation, his conceptions of 'tree' and 'flower' are classified into his conception of 'plant'. Formally speaking, (*Tree* $\sqsubseteq$ *Plant*) $\sqcap$ (*Flower* $\sqsubseteq$ *Plant*). In fact, his conceptions of 'tree' and 'flower' are subsumed under his conception of 'plant'. Semantically, his conceptions of 'tree' and 'flower' are interpretable as his sub-conceptions of 'plant'. Relying on conception classifications, *effect classifications* are analysable. Suppose that Lili's conception of *Learning* is classified into her conceptions of *Writ-*

ing and *Reading*. Subsequently, (*isWriting*(`lili`, `german`) ⊓ *isReading*(`lili`, `german`)) ⊑ *isLearning*(`lili`, `german`). In fact, *isWriting* and *isReading* are the sub-effects of *isLearning* in the case of the singular `german`.

III. Dealing with Conception Characterisation. Conception characterisation characterises the available conceptions of the world. Suppose that Sarah has a conception of 'a (specific) breed of dogs'. This means that she has characterised her conception of *Dog*. It shall be taken into account that the characterisation of a conception is highly dependent on the characterisation of that conception's effects (*effect characterisation*). Correspondingly, Sarah has mainly focused on specifying the properties (as well as the attributes) of that breed of dogs (e.g., having brown colour, being big, being smart). Consequently, she has characterised her conception of *Dog* as well as of 'the effects of being *Dog*'.

IV. Dealing with Conception Identification. Conception identification identifies the available conceptions of the world. James may—based on his conception of *Teacher*—produce his identifying conception of *Teacher*. Equivalently, he may produce his conception of 'being *Teacher*'. It shall be stressed that the identification of a conception is highly dependent on the identification of the properties of that conception as well as of its effects on other conceptions.

V. Dealing with Functionalised Conceptions. Conception functionalisation deals with the functionalisation of the available conceptions of the world. Martin can—based on his conception of 'having *Mother*'—produce his supportive conception that 'any human being has one (and only one) *Mother*'. Obviously, the functionalisation of a conception is supported by the functionalisation of the properties of that conception as well as by the functionalisation of that conception's effects on other conceptions.

VI. Dealing with Nominalised Conceptions. Conception nominalisation nominalises the available conceptions of the world. Robert may—based on his conception of 'having *Rain* in November'—produce his nominalised conception of *Rain*. As another example, Anna has been born on 25 March, 1974. Therefore, she can produce her nominalised conception of *Birth*.

VII. Dealing with Conception Sates. A conception state is a state at which an available conception of the world has been true (has been experienced to be true). More specifically, one—who has experienced a true proposition (like *P*) at a specific state of the world—may decide to develop his conception of *C* based on the experienced *P*. For example, Daniel can—based on his conception of '(many) *Exam*s in February'—make a specific state of his conception of *Exam*. Note that the concept of 'conception state' can have strong junctions with conception nominalisation. As mentioned above, Anna has been born on 25 March, 1974. Therefore, she can—based on her conception of *Birth*—address '25 March, 1974' as a specific state of *Birth* for herself.

VIII. Dealing with Conception Conditioning. Conception conditioning conditions the available conceptions of the world. Juliana can—by qualifying her conception of *Dairy Products*—make her [conditioned] conception of *Worthy Dairy Products*.

IX. Dealing with Conception Comparison. Conception comparison compares the available conceptions of the world with each other. Fred can—based on his conceptions of the singulars `chrome` and `safari` (that are the instances of 'web-browsers')—make his conception comparison.

X. **Dealing with Conception Countability.** By means of conception countability, we can count our available conceptions of the world. Elizabeth can—based on her conception of *Red*—count her red flowers.

7 An Ontology for Conception Representation

This section offers an ontology for conception representation in terminological systems. This $\mathcal{CL}$-based ontology specifies the conceptualisation of (i) humans' conceptions of concepts and (ii) the effects of humans' conceptions of concepts within terminological systems.

7.1 A Formal Ontology for Conception Representation in Terminological Systems

{
NonLogicalSymbol $\sqsubseteq$ *Symbol*,
LogicalSymbol $\sqsubseteq$ *Symbol*,
ConceptionBasedProposition $\sqsubseteq$ *Proposition*,
UnaryPredicate $\sqsubseteq$ *NAryPredicate*,
NAryPredicate $\sqsubseteq$ *Predicate*,
Predicate $\sqsubseteq$ $\exists hasSupport.Proposition$,
NAryPredicate $\sqsubseteq$ *Relation*,
AbstractConception $\sqsubseteq$ *NonLogicalSymbol*,
AbstractConception $\sqsubseteq$ *UnaryPredicate*,
AbstractConception $\sqsubseteq$ $\exists hasInstance.Singular$,
Singular $\sqsubseteq$ *NonLogicalSymbol*,
Singular $\sqsubseteq$ *Constant*,
AbstractEffect $\sqsubseteq$ *NonLogicalSymbol*,
AbstractEffect $\sqsubseteq$ *NAryPredicate*,
AbstractEffect $\sqsubseteq$ $\exists isBasedOn.AbstractConception$,
AbstractEffect $\sqsubseteq$ $\exists hasA.EffectInversion$,
EffectInversion $\sqsubseteq$ *Relation*,
Equivalency $\sqsubseteq$ *Relation*,
Subsumption $\sqsubseteq$ *Relation*,
Classification $\sqsubseteq$ *Relation*,
Opposition $\sqsubseteq$ *Relation*,
Junction $\sqsubseteq$ *Relation*,
Unification $\sqsubseteq$ *Relation*,
Qualification $\sqsubseteq$ *Relation*,

A Formal Ontology for Conception Representation in Terminological Systems 149

Quantification ⊑ *Relation*,
Reflection ⊑ *Relation*,
Transitivity ⊑ *Relation*,
Composition ⊑ *Relation*,
Tautology ⊑ *LogicalSymbol*,
Contradiction ⊑ *LogicalSymbol*,
Tautology ⊑ *ConceptionTruth*,
Contradiction ⊑ *ConceptionFalsity*,
ConceptionTruth ⊑ *ConceptionUniversalQualification*,
ConceptionFalsity ⊑ *ConceptionUniversalQualification*,
ConceptionUniversalQualification ⊑ *ConceptionQualification*,
ConceptionQualification ⊑ *Relation*,
ConceptionQualification ⊑ *ConceptionConditioning*,
ConceptionConditioning ⊑ *Conditioning*,
ConceptionNegation ⊑ *LogicalSymbol*,
ConceptionNegation ⊑ *Relation*,
ConceptionOpposition ⊑ *Opposition*,
ConceptionOpposition ⊑ ∃*hasSupport.ConceptionNegation*,
ConceptionConjunction ⊑ *LogicalSymbol*,
ConceptionConjunction ⊑ *Relation*,
ConceptionJunction ⊑ *Junction*,
ConceptionJunction ⊑ ∃*hasSupport.ConceptionConjunction*,
ConceptionDisjunction ⊑ *LogicalSymbol*,
ConceptionDisjunction ⊑ *Relation*,
ConceptionUnification ⊑ *Unification*,
ConceptionUnification ⊑ ∃*hasSupport.ConceptionDisjunction*,
ConceptionImplication ⊑ *LogicalSymbol*,
ConceptionImplication ⊑ *Relation*,
ConceptionConstruction ⊑ ∃*isBasedOn.AbstractConception*,
ConceptionConstruction ⊑ ∃*hasSupport.LogicalSymbol*,
QualifiedNumberRestriction ⊑ ∃*hasSupport.ConceptionConstruction*,
UnqualifiedNumberRestriction ⊑ ∃*hasSupport.ConceptionConstruction*,
ConceptionFunctionalEffect ⊑ *AbstractEffect*,
ConceptionFunctionalEffect ⊑ ∃*hasSupport.FunctionalConception*,
FunctionalConception ⊑ ∃*hasSupport.FunctionalityOfConception*,
FunctionalityOfConception ⊑ ∃*hasSupport.ConceptionConstruction*,
ConceptionNominalisation ⊑ ∃*isBasedOn.Singular*,
ConceptionNominalisation ⊑ ∃*hasSupport.ConceptionConstruction*,
ConceptionNominalisation ⊑ ∃*hasSupport.OneOfConstruction*,
OneOfConstruction ⊑ ∃*dealWith.ConceptionState*,
OneOfConstruction ⊑
(∃*hasIngredient.ConceptionBasedProposition* ⊓ ∃*hasIngredient.Singular*),
QualifiedNumberRestriction ⊑ *NumberRestriction*,
QualifiedNumberRestriction ⊑ *ConceptionQualification*,
QualifiedNumberRestriction ⊑

($\exists hasIngredient.AbstractConception \sqcap \exists hasIngredient.AbstractEffect$),
$UnqualifiedNumberRestriction \sqsubseteq NumberRestriction$,
$UnqualifiedNumberRestriction \sqsubseteq \exists hasIngredient.AbstractEffect$,
$NumberRestriction \sqsubseteq \exists hasIngredient.NaturalNumber$,
$ConceptionUniversalQuantification \sqsubseteq LogicalSymbol$,
$ConceptionUniversalQuantification \sqsubseteq$
($\exists hasIngredient.AbstractConception \sqcap \exists hasIngredient.AbstractEffect$),
$ConceptionExistentialQuantification \sqsubseteq LogicalSymbol$,
$ConceptionExistentialQuantification \sqsubseteq$
($\exists hasIngredient.AbstractConception \sqcap \exists hasIngredient.AbstractEffect \sqcap \exists hasIngredient.Singular$),
$ConceptionExistentialQuantification \sqsubseteq ConceptionQuantification$,
$ConceptionUniversalQuantification \sqsubseteq ConceptionQuantification$,
$ConceptionQuantification \sqsubseteq Relation$,
$ConceptionQuantification \sqsubseteq ConceptionMeasurement$,
$ConceptionMeasurement \sqsubseteq Measurement$,
$ConceptionMeasurement \sqsubseteq \exists hasProduction.NaturalNumber$,
$ConceptionCharacterisation \sqsubseteq Characterisation$,
$ConceptionCharacterisation \sqsubseteq \exists hasSupport.ConceptionMeasurement$,
$ConceptionComparison \sqsubseteq Comparison$,
$ConceptionComparison \sqsubseteq \exists hasSupport.ConceptionOrdering$,
$ConceptionOrdering \sqsubseteq Ordering$,
$ConceptionOrdering \sqsubseteq \exists hasSupport.NaturalNumber$,
$ConceptionCountability \sqsubseteq Countability$,
$ConceptionCountability \sqsubseteq \exists hasSupport.NaturalNumber$,
$ConceptionIdentification \sqsubseteq \exists hasSupport.ConceptionIdentity$,
$ConceptionIdentity \sqsubseteq Relation$,
$ConceptionIdentity \sqsubseteq (\exists hasIngredient.Identity \sqcap \exists hasIngredient.AbstractConception)$,
$Identity \sqsubseteq AbstractEffect$,
$ConceptionIdentity \sqsubseteq \exists hasSupport.EffectConstruction$,
$ConceptionEquivalency \sqsubseteq Equivalency$,
$ConceptionEquivalency \sqsubseteq ConceptionClassification$,
$ConceptionEquivalency \sqsubseteq TerminologicalConceptionAxiomatisation$,
$ConceptionSubsumption \sqsubseteq Subsumption$,
$ConceptionSubsumption \sqsubseteq ConceptionClassification$,
$ConceptionSubsumption \sqsubseteq TerminologicalConceptionAxiomatisation$,
$ConceptionClassification \sqsubseteq Classification$,
$TerminologicalConceptionAxiomatisation \sqsubseteq TerminologicalAxiomatisation$,
$TerminologicalAxiomatisation \sqsubseteq Axiomatisation$,
$TerminologicalAxiomatisation \sqsubseteq Relation$,
$TerminologicalAxiomatisation \sqsubseteq \exists hasSupport.LogicalSymbol$,
$TerminologicalAxiomatisation \sqsubseteq \exists hasSupport.NonLogicalSymbol$,
$TerminologicalEffectAxiomatisation \sqsubseteq \exists hasSupport.TerminologicalConceptionAxiomatisation$,
$AssertionalAxiomatisation \sqsubseteq Axiomatisation$,
$AssertionalAxiomatisation \sqsubseteq Relation$,
$AssertionalAxiomatisation \sqsubseteq \exists hasSupport.TerminologicalAxiomatisation$,

AssertionalAxiomatisation ⊑ ∃*hasSupport.NonLogicalSymbol*,
ConceptionAssertion ⊑ *AssertionalAxiomatisation*,
ConceptionAssertion ⊑ (∃*hasIngredient.AbstractConception* ⊓ ∃*hasIngredient.Singular*),
EffectAssertion ⊑ *AssertionalAxiomatisation*,
EffectAssertion ⊑ (∃*hasIngredient.AbstractEffect* ⊓ ∃*hasIngredient.Singular*),
EffectJunction ⊑ *Junction*,
EffectJunction ⊑ ∃*hasSupport.EffectConjunction*,
EffectConjunction ⊑ *Relation*,
EffectConjunction ⊑ ∃*hasSupport.EffectConstruction*,
EffectConstruction ⊑ ∃*isBasedOn.AbstractEffect*,
EffectConstruction ⊑ ∃*hasSupport.EffectAssertion*,
EffectUnification ⊑ *Unification*,
EffectUnification ⊑ ∃*hasSupport.EffectDisjunction*,
EffectDisjunction ⊑ *Relation*,
EffectDisjunction ⊑ ∃*hasSupport.EffectConstruction*,
EffectOpposition ⊑ *Opposition*,
EffectOpposition ⊑ ∃*hasSupport.EffectComplement*,
EffectComplement ⊑ *Relation*,
EffectComplement ⊑ ∃*hasSupport.EffectConstruction*,
EffectClassification ⊑ *Classification*,
EffectClassification ⊑ ∃*hasSupport.EffectSubsumption*,
EffectEquivalency ⊑ *Equivalency*,
EffectEquivalency ⊑ *TerminologicalEffectAxiomatisation*,
EffectSubsumption ⊑ *Subsumption*,
EffectSubsumption ⊑ *TerminologicalEffectAxiomatisation*,
EffectComposition ⊑ *Composition*,
EffectComposition ⊑ ∃*hasSupport.EffectReflection*,
EffectReflection ⊑ *Reflection*,
EffectReflection ⊑ ∃*hasSupport.EffectEquivalency*,
EffectComposition ⊑ ∃*hasSupport.EffectEquivalency*,
EffectEquivalency ⊑ *Equivalency*,
EffectComposition ⊑ ∃*hasSupport.EffectTransitivity*,
EffectTransitivity ⊑ *Transitivity*,
EffectTransitivity ⊑ ∃*hasSupport.EffectEquivalency*,
EffectTransitivity ⊑ ∃*hasSupport.EffectSubsumption*,
ComplexEffectSubsumption ⊑ *Relation*,
ComplexEffectSubsumption ⊑ ∃*hasSupport.EffectSubsumption*
}

7.2 A Schematic Ontology for Conception Representation in Terminological Systems

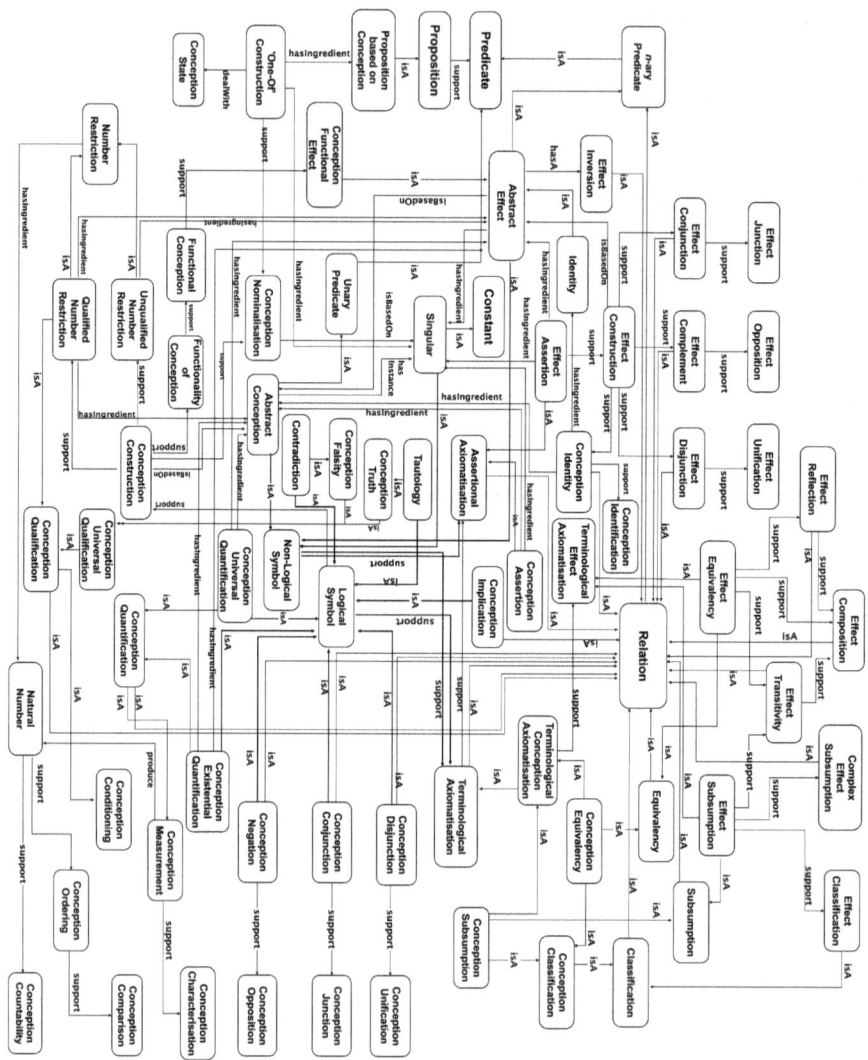

Fig. 1: A $\mathcal{CL}$-based Schematic Ontology for Conception Representation in Terminological Systems. The figure in high resolution is available on http://farshad-badie.com/Resources/

Relying on the described formal ontology, Figure 1 has presented a schematic ontology that conceptually and logically represents the specification of the conceptualisation of humans' conceptions within terminological systems. The schematic ontology is offered based on the following concepts:

1. The 'Logical' and 'Non-Logical' symbols in $\mathcal{CL}$ (that are located at the shallowest layer of the ontology).
2. Terminological and assertional axioms in $\mathcal{CL}$.
3. The constructed logical conceptions in $\mathcal{CL}$.
4. The constructed logical effects in $\mathcal{CL}$.
5. The most fundamental logical concepts in connection to logical conceptions/effects.
6. The concepts of 'Constant', 'Relation' and 'Predicate'. In fact, as pointed out in the paper, $\mathcal{CL}$ is a fragment of predicate logic. Correspondingly, our ontology has reached these concepts.
7. The concept of 'Proposition' is located at the deepest layer of the ontology.

8 Conclusion

Concepts (as conceptual entities) are the basic materials of meanings (that are conceptual structures) and are the primary and fundamental units of humans' terminological knowledge. This research has taken into account that human beings—for expressing their own world descriptions based on their mental concepts—become concerned with linguistic expressions (that are expressed based on, e.g., letters, symbols). Hence, they produce their own 'conceptions' of the world. Thereby, in my theoretical model, conceptions are the consequences and the productions of humans' mental concepts. From the perspective of formal logics, conceptions (of concepts) generate hypotheses in formal-logical systems. For the logical assessment of conceptions, any conception and its interconnections with other conceptions can—by means of predication—be assigned into logical symbols.

It has been the underlying research hypothesis of this paper that the most fundamental building blocks of logical-terminological descriptions are expressible based on humans' conceptions. Hence it has been taken into consideration that we need a formal system to represent and explicate humans' conceptions. According to this paper, such a formal system is representable based on a description logic like Conception Language ($\mathcal{CL}$). $\mathcal{CL}$ is a decidable fragment of predicate logic and represents knowledge in terms of (i) conceptions, (ii) conceptions' instances, and (ii) conceptions' effects (on other conceptions).

As the most important contribution, the paper has offered a formal (and, accordingly, a schematic ontology) for conception representation in terminological systems. The ontology has specified the conceptualisation of humans' conceptions as well as of the effects of their conceptions.

References

A HJ, E Moss H (2003) Concepts and meaning: Introduction to the special issue on conceptual representation, language and cognitive processes. Taylor & Francis

Allwood J (1999) On concepts, their determination, analysis and construction. In: Gothenburg Papers in Theoretical Linguistics, Department of Linguistics, University of Gothenburg

Allwood J, Andersson LG (1976) Semantik, GULING 1. Department of Linguistics, University of Gothenburg

Baader F, Calvanese D, McGuinness DL, Nardi D, Patel-Schneider PF (2010) The Description Logic Handbook: Theory, Implementation and Applications, 2nd edn. Cambridge University Press, New York, USA

Baader Fea (2017) An Introduction to Description Logic. Cambridge University Press

Badie F (2017a) From concepts to predicates within constructivist epistemology. In: Baltag A, Seligman J, Yamada T (eds) LORI, Springer, Lecture Notes in Computer Science, vol 10455, pp 687–692

Badie F (2017b) A theoretical model for meaning construction through constructivist concept learning: A conceptual, terminological, logical and semantic study within human-human-machine interactions. PhD thesis, Aalborg Universitet, DOI 10.5278/vbn.phd.hum.00073

Bartlett FC (1932) A study in experimental and social psychology. Cambridge University Press

Brachman RJ, Levesque HJ (eds) (1985) Readings in Knowledge Representation. Morgan Kaufmann

Buchheit M, Donini FM, Schaerf A (1993) Decidable reasoning in terminological knowledge representation systems. CoRR cs.AI/9312101

van Harmelen F, Lifschitz V, Porter B (eds) (2008) Handbook of Knowledge Representation. Foundations of Artificial Intelligence, Elsevier Science

Kant I (1781/1964) Critique of pure reason. Macmillan, London

Levesque HJ, Brachman RJ (1985) A fundamental tradeoff in knowledge representation and reasoning. Readings in Knowledge Representation, Morgan Kaufmann

Margolis E, Laurence S (2007) The ontology of concepts — abstract objects or mental representations ? NOÛS 41(4):561–593

Margolis E, Laurence S (2010) Concepts and theoretical unification. The Behavioral and brain sciences 33 2-3:219–20

Margolis S E & Laurence (2015) The Conceptual Mind: New Directions in the Study of Concepts. MIT Press

Minsky M (1974) A framework for representing knowledge. Tech. Rep. 306, MIT-AI Laboratory

Patrick Blackburn FW Johan van Benthem (ed) (2006) Handbook of Modal Logic, Studies in Logic and Practical Reasoning, vol 3. Elsevier Science

Peacocke C (1992) A Study of Concepts. Cambridge, MA: MIT Press

Quillian R (1968) Semantic memory. In: Semantic Information Processing, MIT Press, pp 216–270

Rudolph S (2011) Foundations of description logics. In: In Reasoning Web, volume 6848 of LNCS, Springer

Sikos LF (2017) Description Logics in Multimedia Reasoning. Springer International Publishing, DOI 10.1007/978-3-319-54066-5

What is refutation?

Gabriele Pulcini* and Tomasz Skura

Abstract The paper offers a logical analysis of the concept of refutation and illustrates some possible directions of research in the field of philosophical logic as well as in the methodology of propositional calculi.

Key words: refutation rules, axiomatic refutation systems, methodology of propositional logic, philosophy of logic, meaning theory

1 Introduction

We analyze the concept of refutation in both philosophical and mathematical logic. First, we deal with axiomatic refutation systems. Two types of such systems are considered. Systems with *reverse substitution* (having theoretical applications) and systems without this rule (applicable in counter-model constructions and decision procedures). Then, we consider a list of topics, most of them in philosophical logic, which could benefit from the specific technical and conceptual tools offered by the refutational approach.

Gabriele Pulcini
Institute for Logic, Language and Computation
University of Amsterdam
e-mail: g.pulcini@uva.nl

Tomasz Skura
Institute of Philosophy
University of Zielona Góra
e-mail: t.skura@ifil.uz.zgora.pl

* The first author thankfully acknowledges the support from the Dutch Research Council (NWO) through the Open Competition-SSH project 406.18.TW.009 "A Sentence Uttered Makes a World Appear—Natural Language Interpretation as Abductive Model Generation".

2 Axiomatic Refutation Systems

In formal logic, one can find two approaches to refutation: an indirect one and a direct one. In the indirect approach, you refute a formula by failing to prove it. For example, you search for a proof of A, and if all the possibilities of finding a proof for A have been exhausted, you say that A is refuted. In the direct approach, a single refutation of A, which is a derivation, justifies refuting A.

Although the indirect approach is standard, we believe that combining two direct approaches (one looking for a proof and the other for a refutation) is more attractive and can yield new results that are both interesting and useful. Note that if A is non-valid, then finding a refutation for A may be simpler than producing all possibilities of proving it.

2.1 Basic Concepts

An axiomatic refutation system is just like a traditional axiomatic system, but it is applied to non-valid formulas rather than valid ones.

Let **L** be a logic, that is, a set of formulas closed under *substitution*, *modus ponens*, and possibly some other rules (e.g. *necessitation*). We say that a rule

$$\frac{A_1, \ldots, A_n}{B}$$

is a *refutation rule for* **L** iff $B \notin \mathbf{L}$ whenever every $A_i \notin \mathbf{L}$. A *refutation system for* **L** is a pair $\mathbf{S} = (AX, RU)$, where AX is a set of formulas that are not in **L** (called refutation axioms for **L**) and RU is a set of refutation rules for **L**. A formula A is **S**-*refutable* (in symbols $\dashv_\mathbf{S} A$, or just $\dashv A$) iff A has a derivation in **S**. Moreover, we say that a refutation system **S** is *characteristic for* **L** iff, for every formula A, we have:

$$A \notin \mathbf{L} \text{ iff } \dashv_\mathbf{S} A.$$

In the literature, various kinds of refutation rules can be found. Here, we focus on refutation rules preserving non-validity (for another approach, see e.g. Fiorentini and Ferrari 2017). Let us start with the following refutation rules introduced by (Łukasiewicz, 1951).

- *Reverse substitution (RS)*: B/A where B is a substitution instance of A.
- *Reverse modus ponens (RMP)*: B/A where $\vdash A \rightarrow B$.

Remark 1. In this section, by a rule we mean a set of pairs Γ/A, where $\Gamma \cup \{A\}$ is a set of formulas. So, our description of RMP fits into the above definition of a refutation rule.

Remark 2. We sometimes present refutation rules bottom-up:

$$\frac{B}{A_1 \mid ... \mid A_n}$$

as multiple-conclusion rules preserving validity. (If $B \in \mathbf{L}$ then some $A_i \in \mathbf{L}$).

2.2 Systems with RS

CL (Classical Propositional Logic) can be characterized by the following simple refutation system.

Refutation axiom: $\bot$ (the false).
Refutation rules: RS, RMP.

Refutation systems with *RS* (involving certain characteristic formulas of finite algebras as refutation axioms) characterize every intermediate logic (and every normal modal logic) with the *FMP* (finite model property) (Skura, 1992; Skura, 1994, 2013; Citkin, 2013). However, there are logics without the *FMP* (that is, they cannot be characterized by any class of finite models) that do have finite refutation systems (Skura, 1992; Skura, 1994, 2013).

Furthermore, there are logics with problematic (or unknown) proof theories, but having neat syntactic descriptions of their non-validities. For example, Medvedev's logic is characterized by the following refutation system (Skura, 1992).

Refutation axiom: $\bot$ (the false).
Refutation rules: RS, RMP_{KP}, RD, where

(RMP_{KP}) B/A where $\vdash_{KP} A \to B$.

Here $\vdash_{KP}$ means provability in the Kreisel-Putnam logic, which is the extension of Intuitionistic Logic (**Int**) by the axiom

$$(\neg A \to B \vee C) \to (\neg A \to B) \vee (\neg A \to C).$$

(RD) $A, B/A \vee B$ (This rule reverses the disjunction property.)

Also, refutation systems with *RS* are useful for establishing certain facts about the lattice of extensions of a given logic, especially, concerning maximality and minimality (Skura, 2004, 2009).

Another duality concerns the way you create a non-classical logic:

- (Positive approach) You reject some unacceptable classical law (for example, the law of explosion or the positive paradox) and derive provable formulas from the acceptable axioms getting **P** (Paraconsistent Logic or Implicational Relevance Logic).
- (Negative approach) You keep the rejected law as a refutation axiom and you declare a formula *refutable* iff the refutation axiom is derivable from it. The set of refutable formulas is thus defined. If the complement **N** of this set is closed under the inference rules, then **N** is our new logic (Skura, 2004, 2017a).

Humorously speaking, in the positive approach, we want what is good, and in the negative approach, we prevent what is bad. **P** and **N** are the two extremes of possible solutions.

2.3 Systems without RS

Constructing Counter-Models

Of course, *reverse substitution* is not good for constructing counter-models. However, *reverse modus ponens* is okay, but in Johansson's logic and extensions (including **Int** and intermediate logics), it must have the following form.

$(RMP')\ A \to C / B \to C$ where $\vdash A \to B$.

Roughly speaking, in Johansson's logic and extensions as well as in normal modal logics, finite countermodels can be constructed from syntactic refutations (which can be presented as finite trees consisting of formulas) as follows.

- For every formula A, we construct its Mints-normal form F_A such that $\vdash A$ iff $\vdash F_A$ (Mints, 1990). Every normal form has its natural number called its rank (Skura, 2011, 2013).
- We give a Scott-style refutation rule involving normal forms.

  $(R) \quad \dfrac{F_1, ..., F_n}{F}$

  where F is a normal form of rank $n > 0$ and each F_i is (after simple *modus ponens* transformations) a normal form of rank smaller than n (Scott, 1957; Skura, 2011, 2013).
- Our refutation system consists of refutation axioms (which are normal forms of rank 0 that are non-valid) and refutation rules: R and RMP (or RMP').
- We prove, by a simple inductive argument, that every normal form F is either provable or refutable (Scott, 1957; Skura, 2011, 2013). So, if F is not provable then F has a refutation tree.
- We transform every syntactic refutation tree into a Kripke frame by removing the nodes obtained by *RMP* and by defining a suitable accessibility relation (Skura, 2002, 2013, 2017a). From the normal forms, we extract a valuation falsifying the refutable formulas. So, if F is a node in a syntactic refutation tree, then it is false at some point in the corresponding model built from this tree. (We remark that the corresponding frames need not be trees.) Hence if F is refutable then F has a countermodel, so F is not provable.
- As a result we get both syntactic completeness (F is not provable iff F is refutable) and semantic completeness (F is provable iff F is valid in all finite tree-type frames).

Decision Procedures

Of course, *RMP* is not good for refutation search procedures. However, we do not need the whole *RMP* in our syntactic completeness proof (Skura, 2011). Just a few simple auxiliary rules are enough. The completeness proof provides a refutation search procedure that is a finite tree consisting of finite sets of formulas and having the following property: the origin is non-valid iff some end node is non-valid (Skura, 2017b). (Here we say that a set of formulas is non-valid iff every member of it is non-valid.) Note that it is in fact a decision procedure.

Refutation Procedures and Tableau Procedures

We focus on Modal Logic and follow Goré in our account of tableau systems (Goré, 1999). (We assume the reader to be familiar with basic concepts concerning tableaux.)

Tableau procedures are viewed as refutation procedures in the sense that in order to show that a formula *A* is valid we assume that *A* is not valid (Fitting, 1999). Then we apply tableau rules

$$\frac{\Gamma}{\Gamma_1 \mid ... \mid \Gamma_k}$$

where $\Gamma, \Gamma_1, .., \Gamma_k$ are finite sets of formulas. The interpretation is that if Γ is satisfiable, then so is some Γ_i. As a result, we get a tableau for $\{\neg A\}$, which is a finite tree with origin $\{\neg A\}$. A formula *A* is *tableau-provable* iff there is a closed tableau for $\{\neg A\}$. Tableau procedures are a generalization of disjunctive normal form procedures.

It is easier to compare refutation procedures with tableau procedures when refutation rules are presented bottom-up (see Remark 2). In a bottom-up refutation procedure, we assume that *A* is valid. Then we apply bottom-up refutation rules generating finite trees consisting of formulas. A formula *A* is refutable iff there is a finite refutation tree with origin *A* and refutation axioms as end-nodes. Refutation procedures are a generalization of conjunctive normal form procedures. Thus, as syntactic procedures, tableau procedures and refutation procedures are complementary.

Tableau procedures also provide counter-model constructions. Assume that there is no closed tableau for $\{\neg A\}$. Then a finite tree-type counter-model for *A* can be constructed from various open tableaux for $\{\neg A\}$. Note that in transitive logics (**K4** and extensions) such constructions involve cycles.

On the other hand, we get a finite tree-type counter-model for *A* from a refutation tree for *A directly*, by deleting the nodes obtained by *RMP* and transforming the resulting tree into a frame (Skura, 2002, 2013). We remark that our constructions are cycle-free. It turns out that, at least in some cases, our counter-model constructions are simpler than the tableau constructions (Skura, 2013, p.125).

3 Concluding remarks and research perspectives

In short, philosophical logic is intended to provide the interface through which philosophy and logic interact. On the one hand, philosophical logic concerns the study of philosophical problems — especially problems in epistemology, (analytic) metaphysics and the philosophy of language — by means of the specific mathematical tools offered by logic (Grayling, 1982). On the other hand, once logically addressed, meaningful philosophical settings never fail to produce challenging new logical problems to be solved mathematically.

In what follows we provide a list of topics in philosophical logic and, more in general, in the methodology of propositional logic which could benefit from the specific formal and conceptual tools made available by refutation systems.

3.1 *General proof-theory program*

In 1974, Prawitz proposed to emancipate proof-theory from classical foundational studies by means of a new research program that he termed *general proof-theory*. The core of this program consists in the fact that "proofs are studied in their own right where one is interested in general questions about the nature and structure of proofs [...]" (Prawitz, 1974, p. 66).

The introduction of refutation calculi can be seen as a way to 'maximize' Prawitz's program inasmuch as these systems allow us to widen the space of proofs by including derivations ending with *invalid* formulas/sequents (Varzi, 1990, 1992; Goranko, 1994; Skura, 2009, 2011b; Piazza and Pulcini, 2016; Carnielli and Pulcini, 2017; Piazza and Pulcini, 2019). Here the basic idea is that one can have a better understanding of the structure of proofs once the 'affirmative' and the 'refutational' parts are considered together as two alternative and complementary ways to syntactically characterize a given (decidable) logic.

On the one hand, this idea echoes what happens in Girard's *Ludics*, where proofs and para-proofs peacefully co-exist and interact (Girard, 2001). On the other, it can be also seen as a way to further extend Wansing's proposal of accommodating dual proofs logically (Wansing, 2017). Indeed, from an epistemic point of view, ontological parsimony is more an obstacle to the real understanding of the structure of proofs than a virtue.

3.2 *Meaning theory and proof-theoretic semantics*

As a byproduct of the ontological extension posited in the previous point, refutation calculi might provide new conceptual tools in the fields of anti-realist meaning theory and proof-theoretic semantics (Dummett, 1975; Schroeder-Heister, 2018).

Hardcore proof-theorists tend to believe that the meaning of logical operators is primarily conveyed by their rules in a suitable proof-system. Model-theoretic semantics comes into play at a later moment to provide a mathematical account of these very epistemic insights (Sundholm, 1986; Schroeder-Heister, 2018; Kremer, 1988). Yet this view has been discussed and criticized on several occasions; in this regard, the logico-philosophical debate about the *tonk* connective is very well-known (Prior, 1960; Belnap, 1962; Avron, 2010).

It is our conviction that several of the problems arising in this field might be fruitfully addressed by considering the meaning of logical operators as given by their rules in *both* the affirmative and the refutational parts. Take, for instance, the classical conjunction operator ($\wedge$). According to this view, its meaning should be considered as being conveyed not only through its rules in the affirmative part LK:

$$\frac{\Gamma, A, B \vdash \Delta}{\Gamma, A \wedge B \vdash \Delta} \wedge \vdash \qquad \frac{\Gamma \vdash A, \Delta \quad \Gamma' \vdash B, \Delta'}{\Gamma, \Gamma' \vdash A \wedge B, \Delta, \Delta'} \vdash \wedge,$$

but also by its rules in the refutational part $\overline{\text{LK}}$ (Goranko, 1994):

$$\frac{\Gamma, A, B \dashv \Delta}{\Gamma, A \wedge B \dashv \Delta} \wedge \dashv \qquad \frac{\Gamma \dashv A, \Delta}{\Gamma \dashv A \wedge B, \Delta} \dashv \wedge(1) \qquad \frac{\Gamma \dashv B, \Delta}{\Gamma \dashv A \wedge B, \Delta} \dashv \wedge(2).$$

It turns out that negation ($\neg$) is the only classical connective which allows for the very same (left and right) introduction rules in both parts, affirmative and negative:

$$\frac{\Gamma \vdash \Delta, A}{\Gamma, \neg A \vdash \Delta} \neg \vdash \qquad \frac{\Gamma, A \vdash \Delta}{\Gamma \vdash \Delta, \neg A} \vdash \neg$$

$$\frac{\Gamma \dashv \Delta, A}{\Gamma, \neg A \dashv \Delta} \neg \dashv \qquad \frac{\Gamma, A \dashv \Delta}{\Gamma \vdash \Delta, \neg A} \dashv \neg$$

3.3 Comparative theory of formalisms

Formalized proofs and the study of their structural properties always relate to a specific proof-system of reference, and new formalisms are expected to improve the already known deductive engines in some respects (proof-search effectiveness, naturalness, perspicuity, identity of proofs, etc.). Put thus, besides being a theory of proofs, proof-theory can be also conceived of as a comparative meta-theory of formalisms. In this regard, refutation calculi might suggest new strategies for improving our proof-systems and challenging new problems. To take an example, the problem of designing a satisfactory proof-net theory for classical logic still remains, in many respects, an open problem (Girard, 1987). Surprisingly enough, once refutationally addressed, classical logic comes with a very simple and efficient proof-net theory which may, in turn, provide new information about the deeper geometrical structure of proof-nets for the affirmative counterpart (Pulcini and Varzi, 2019).

3.4 Philosophy of logic

According to (Carnielli and Pulcini, 2017), we indicate with $\overline{\mathsf{LK}}$ the sequent system sound and complete with respect to the set of classically invalid sequents (Goranko, 1994). It seems quite reasonable to assert that $\overline{\mathsf{LK}}$ is actually a *logic*, to the extent that it provides an alternative syntactic characterization of propositional classical logic (Casati and Varzi, 2000; Pulcini and Varzi, 2018).

It is worth observing that, unlike LK, $\overline{\mathsf{LK}}$ is *paraconsistent*. In other words, classical logic can be syntactically grasped, albeit in the negative, by means of a paraconsistent sequent system. In general, any decidable logic whose semantics circumscribes a set of contingent formulas allows for a refutational characterization which is paraconsistent (Pulcini and Varzi, 2018). This kind of observation offers new insights into the logical nature of paraconsistency, which turns out to be sensitive to the specific syntactic formulation of our logic. The system $\overline{\mathsf{LK}}$ does not need to resort to structural rules either, therefore similar considerations also apply to the notion of *substructurality* (Restall, 2018).

3.5 Rejection / assertion debate

Is the negation of a proposition A the same as the *rejection* of A? Frege famously maintained that rejection and assertion do not have to be treated separately, since the act of rejecting a proposition A is nothing but the act of affirming its negation (Frege, 1919). The opposite view is called *bilateralism* and finds in Smiley one of its more tenacious proponents (Smiley, 1996).

This debate has remained lively and intense over the years, involving both philosophers of language, linguists, and logicians interested in investigating the nature of the negation operator (Rumfitt, 2000; Incurvati and Smith, 2010; Ripley, 2011; Incurvati and Schlöder, 2017). It would be interesting to understand how technical advances in the proof-theory of rejection calculi might contribute to the elucidation of this rejection/negation rapport.

3.6 Metaphysical grounding

In metaphysics, the ground relation is meant to connect the truth of some set of facts $A_1, \ldots, A_n$ to the truth of some other fact B. This relation has to be something conceptually deeper than mere logical implication: it has to explain *why* B is true, by virtue of the truth of each one of the As. Put differently, when we are in presence of a ground relation, it is *metaphysically necessary* that the truth of the fact B came from the truth of the premises $A_1, \ldots, A_n$ (Fine, 2012).

Logic enters the debate at the moment we want to grasp metaphysical necessitation by means of a suitable set of formal constraints regulating the transition from $A_1, \ldots, A_n$

to B. Needless to say, the resort to formal logic has been usually intended as the resort to the affirmative part of logical systems (Fine, 2012). We guess that refutation systems could provide new tools to fine-tune the formal grasp of grounding by allowing us to introduce complementary considerations about the deductive transmission of *un*grounding.

3.7 Methodology of propositional logics

Our analysis shows that the concept of refutation provides new tools having the following interesting applications in the methodology of propositional logics.

- Specific/generic descriptions of the non-validities of logics.
- Establishing maximality/minimality in the lattices of logics.
- Non-classical logics *via* refutability.
- Constructive completeness proofs that are simple.
- Refined semantic characterizations of logics by finite tree-type frames.
- Cycle-free constructions of counter-models.
- Refutation search procedures that, at least in some cases, are simpler than those provided by standard methods.

References

Avron A (2010) Tonk—A full mathematical solution. Ruth Manor's Festschrift
Belnap N (1962) Tonk, plonk and plink. Analysis 22(6):130–134
Carnielli WA, Pulcini G (2017) Cut-elimination and deductive polarization in complementary classical logic. Logic Journal of the IGPL 25(3):273–282
Casati R, Varzi AC (2000) True and false: An exchange
Citkin A (2013) Jankov-style formulas and refutation systems. Reports on Mathematical Logic 48:67–80
Dummett MAE (1975) What is a theory of meaning? In: Guttenplan S (ed) Mind and Language, Oxford University Press
Fine K (2012) Guide to ground. Metaphysical grounding: Understanding the structure of reality pp 37–80
Fiorentini C, Ferrari M (2017) A forward unprovability calculus for Intuitionistic Propositional Logic. In: TABLEAUX 2017, Springer, pp 114–130
Fitting M (1999) Introduction. In: Agostino MD, Gabbay D, Hähnle R, Posega J (eds) Handbook of Tableau Methods, Kluwer, pp 1–43
Frege G (1919) Die verneinung. eine logische untersuchung. Beiträge zur Philosophie des Deutschen Idealismus I 3-4:143–157
Girard J-Y (1987) Linear logic. Theor Comput Sci 50:1–102, DOI 10.1016/0304-3975(87)90045-4, URL http://dx.doi.org/10.1016/0304-3975(87)90045-4

Girard J-Y (2001) Locus solum: From the rules of logic to the logic of rules. Mathematical Structures in Computer Science 11(3):301–506, DOI 10.1017/S096012950100336X, URL https://doi.org/10.1017/S096012950100336X

Goranko V (1994) Refutation systems in modal logic. Studia Logica 53(2):299–324, DOI 10.1007/BF01054714, URL http://dx.doi.org/10.1007/BF01054714

Goré R (1999) Tableau methods for modal and temporal logics. In: Agostino MD, Gabbay D, Hähnle R, Posega J (eds) Handbook of Tableau Methods, Kluwer, pp 297–396

Grayling AC (1982) An Introduction to Philosophical Logic. Harvester studies in philosophy, Harverster Press, URL https://books.google.it/books?id=dXyuAAAAIAAJ

Incurvati L, Schlöder JJ (2017) Weak rejection. Australasian Journal of Philosophy 95(4):741–760, DOI 10.1080/00048402.2016.1277771

Incurvati L, Smith P (2010) Rejection and valuations. Analysis 70(1):3–10

Kremer M (1988) Logic and meaning: The philosophical significance of the sequent calculus. Mind 97(385):50–72

Łukasiewicz J (1951) Aristotle's Syllogistic from the Standpoint of Modern formal Logic. Clarendon, Oxford

Mints G (1990) Gentzen-type systems and resolution rules. Lecture Notes in Computer Science 417:198–231

Piazza M, Pulcini G (2019) Fractional semantics for classical logic. The Review of Symbolic Logic pp 1–19, DOI 10.1017/S1755020319000431

Piazza M, Pulcini G (2016) Uniqueness of axiomatic extensions of cut-free classical propositional logic. Logic Journal of the IGPL 24(5):708–718, DOI 10.1093/jigpal/jzw032, URL http://dx.doi.org/10.1093/jigpal/jzw032

Prawitz D (1974) On the idea of a general proof theory. Synthese 27(1-2):63–77

Prior AN (1960) The runabout inference-ticket. Analysis 21(2):38–39

Pulcini G, Varzi AC (2018) Paraconsistency in classical logic. Synthese 195(12):5485–5496

Pulcini G, Varzi AC (2019) Proof-nets for non-theorems, unpublished

Restall G (2018) Substructural logics. In: Zalta EN (ed) The Stanford Encyclopedia of Philosophy, spring 2018 edn, Metaphysics Research Lab, Stanford University

Ripley D (2011) Negation, denial, and rejection. Philosophy Compass 6(9):622–629

Rumfitt I (2000) 'yes and no'. Mind 109(436):781–823

Schroeder-Heister P (2018) Proof-theoretic semantics. In: Zalta EN (ed) The Stanford Encyclopedia of Philosophy, spring 2018 edn, Metaphysics Research Lab, Stanford University

Scott D (1957) Completeness proofs for the intuitionistic sentential calculus. In: Summaries of talks presented at the Summer Institute of Symbolic Logic, Cornell University, second edition, Princeton, 1960, 231-241

Skura T (1992) Refutation calculi for certain intermediate propositional logics. Notre Dame Journal of Formal Logic 33:552–560

Skura T (1994) Syntactic refutations against finite models in modal logic. Notre Dame Journal of Formal Logic 35(4):595–605, DOI 10.1305/ndjfl/1040408615, URL https://doi.org/10.1305/ndjfl/1040408615

Skura T (2002) Refutations, proofs, and models in the modal logic K4. Studia Logica 70(2):193–204, DOI 10.1023/A:1015174332202, URL https://doi.org/10.1023/A:1015174332202

Skura T (2004) Maximality and refutability. Notre Dame Journal of Formal Logic 45:65–72

Skura T (2009) A refutation theory. Logica Universalis 3(2):293–302, DOI 10.1007/s11787-009-0009-y, URL http://dx.doi.org/10.1007/s11787-009-0009-y

Skura T (2011a) Refutation systems in propositional logic. In: Gabbay DM, Guenthner F (eds) Handbook of Philosophical Logic, vol 16, Springer, pp 115–157

Skura T (2013) Refutation Methods in Modal Propositional Logic. Semper, Warszawa

Skura T (2017a) The greatest paraconsistent analogue of Intuitionistic Logic. In: Proceedings of the 11th Panhellenic Logic Symposium, pp 71–76

Skura T (2017b) Refutations in Wansing's logic. Reports on Mathematical Logic 52:83–99

Skura T (2011b) On refutation rules. Logica Universalis 5(2):249–254, DOI 10.1007/s11787-011-0035-4, URL https://doi.org/10.1007/s11787-011-0035-4

Smiley T (1996) Rejection. Analysis 56(1):1–9

Sundholm G (1986) Proof theory and meaning. In: Handbook of philosophical logic, Springer, pp 471–506

Varzi AC (1990) Complementary sentential logics. Bulletin of the Section of Logic 19(4):112–116

Varzi AC (1992) Complementary logics for classical propositional languages. Kriterion Zeitschrift für Philosophie 4:20–24

Wansing H (2017) A more general general proof theory. J Applied Logic 25:23–46, DOI 10.1016/j.jal.2017.01.002, URL https://doi.org/10.1016/j.jal.2017.01.002

Refutation systems in the finite

Valentin Goranko and Tomasz Skura

Abstract Refutation systems are deductive systems intended to derive the non-valid, i.e. (semantically) refutable formulae of a given logical system. The goal of this paper is to present some refutation systems on finite semantic structures and establish some basic facts about them. In particular, we develop generic refutation systems for modal logics and for first-order theories that are semantically determined by single finite structures or by classes of finite structures, for arbitrary first-order languages.

Key words: refutation system, refutability in the finite, modal logic, first-order logic

1 Introduction

Refutation systems are deductive systems intended to derive the non-valid, i.e. (semantically) refutable formulae of a given logical system. A (syntactic) refutation system R typically consists of refutation axioms and of refutation rules. We say that a formula φ is *refutable in* R, usually denoted by R $\dashv \varphi$, iff it is derivable in R from the refutation axioms by using the refutation rules. For more on refutation systems see e.g. Skura (2011).

In this paper we develop generic refutation systems for modal logics and first-order theories that are semantically determined by single finite structures or classes of finite structures, for arbitrary first-order languages. We prove that these systems are sound

---

Valentin Goranko
Stockholm University, Sweden
University of Johannesburg (visiting professorship)
e-mail: valentin.goranko@philosophy.su.se

Tomasz Skura
University of Zielona Góra, Poland
e-mail: T.Skura@ifil.uz.zgora.pl

and refutation-complete, i.e. they derive all, and only, non-valid formulae in their respective semantics.

Generally, such results could be regarded just as technical exercises in translating semantic into syntactic refutations, but they do have their own value. A good example is FO in the finite FO^{fin}, which, by Trakhtenbrot's theorem (see e.g. Ebbinghaus and Flum (2005)), has no recursive axiomatization. However, the set of non-valid formulae of FO^{fin} is recursively enumerable. Therefore, they can, in principle, be axiomatized by refutation systems, and such systems have both technical and methodological importance. We briefly discuss that issue further in the paper.

Some related works on refutation systems in the finite include:

- In Hailperin (1961) probably the first explicitly developed axiomatic refutation system for FO, for languages without equality and function symbols, was presented and proved sound and complete for falsifiability in finite FO models.
- In Tiomkin (1988) a Gentzen type refutation system with a natural set of rules for FO with equality but again without function symbols, is proposed for deriving falsifiable sequents ('antisequents'). That system was proved sound and complete for falsifiability in finite FO models, and an application for loop detecting in Prolog programs is proposed in that paper.
- Some general results about refutation systems for for modal logics corresponding to (classes) of finite models were obtained in Goranko (1994), Skura (1994), Skura (2013), Citkin (2013). The latter three of these papers deal with modal algebras, whereas the first one is based on Kripke frames.

2 Preliminaries

2.1 Basic logical notation and terminology

We assume that the reader has basic background in modal logic, including Kripke frames and models, Kripke semantics, incl. satisfiability and validity of modal formulae, and bisimulations and n-bisimulations between Kripke models. For these, refer e.g. to Goranko and Otto (2007). We only provide here basic notation and terminology of modal logic used further.

We denote the classical propositional logic by PL. Let ML be the standard modal propositional logic extending PL, with formulae generated from a set of atomic propositions $AT = \{p_1, p_2, ...\}$ (for which we use p, q as metavariables) by using the constant $\bot$ and the connectives $\neg, \wedge, \vee, \rightarrow, \Box$. We also define, as usual, $\top := \neg\bot$, $\varphi \leftrightarrow \psi := (\varphi \rightarrow \psi) \wedge (\psi \rightarrow \varphi)$ and $\Diamond \varphi := \neg \Box \neg \varphi$. If Φ, Ψ are finite sets of formulae, then $\Phi \longrightarrow \Psi$ stands for $\bigwedge \Phi \rightarrow \bigvee \Psi$. We put, as usual, $\bigvee \varnothing := \bot$ and $\bigwedge \varnothing := \top$. We also write Φ, Ψ for $\Phi \cup \Psi$. A *(Kripke) frame* is a pair $\mathcal{F} = \langle W, R \rangle$, where W is a nonempty set of points (worlds) and R is a binary relation on W. A *(Kripke) model* is a pair $\langle \mathcal{F}, V \rangle$, where $\mathcal{F}$ is a frame and $V : VAR \mapsto \mathcal{P}(W)$ is a *valuation*, extended to all formulae in the standard way. We denote truth of a modal formula φ at a world w in a Kripke model $\langle \mathcal{F}, V \rangle$ by $\langle \mathcal{F}, V \rangle, w \models \varphi$. The formula φ is *valid in* $\langle \mathcal{F}, V \rangle$, denoted $\langle \mathcal{F}, V \rangle \models \varphi$, iff $\langle \mathcal{F}, V \rangle, w \models \varphi$ for each $w \in \mathcal{F}$; φ is *valid in* $\mathcal{F}$, denoted $\mathcal{F} \models \varphi$, iff $\langle \mathcal{F}, V \rangle \models \varphi$ for every valuation V. Finally, φ is *valid*, denoted $\models \varphi$, iff it is valid in every frame.

We will also be working with first-order languages with any (finite) signature, containing relational, constant and functional symbols. Let us fix any such language L.

- First-order terms and formulae in L are defined as usual. We will denote constants by $a, b, c, d, ...$, variables by $u, v, x, y, ...$, and terms by s, t – all possibly with indices.
- An *atomic formula* is either $t_1 = t_2$, where t_1, t_2 are terms, or $Rt_1, ..., t_n$, where R is an *n*-ary relation symbol and $t_1, ..., t_n$ are terms.
- The set of FO formulae *FOR* is generated from the atomic formulae by the connectives $\rightarrow, \wedge, \vee, \neg$ and the quantifiers $\forall, \exists$. A *literal* is an atomic formula or a negated one. We write $t_1 \neq t_2$ for $\neg t_1 = t_2$.

Let $\varphi = \varphi(x_1, ..., x_n)$ be any FO formula in the language L with free variables amongst $x_1, ..., x_n$, let $\mathcal{A}$ be an FO structure for L, and let α be an assignment in $\mathcal{A}$ such that $\alpha(x_1) = a_1, ..., \alpha(x_n) = a_n$. Then we write $\mathcal{A} \models \varphi[a_1, ..., a_n]$ for $\mathcal{A} \models \varphi[\alpha]$. If φ is a sentence, we write simply $\mathcal{A} \models \varphi$ (read: "$\mathcal{A}$ satisfies φ") instead of "$\mathcal{A} \models \varphi[\alpha]$ for some (hence, every) α".

For other standard concepts and notation of FO used here see e.g. in Ebbinghaus and Flum (2005).

Let us fix any complete standard axiomatic system Ax$_{\text{FO}}$ for FO (see e.g. Barwise (1977) or Goranko (2016)). A formula φ is Ax$_{\text{FO}}$-*provable* iff φ has a derivation in Ax$_{\text{FO}}$. By completeness, φ is Ax$_{\text{FO}}$-*provable* iff φ is true in all structures under all assignments.

2.2 Refutation systems: basic concepts

We now define basic concepts by modifying and extending some terminology and results from Goranko (1994), Goré and Postniece (2008). In general, by a sequent we mean $\Phi \bowtie \Psi$, where $\bowtie \in \{\vdash, \dashv\}$. However, here we will only be dealing with sequents $\bowtie \varphi$, that is, sequents with $\Phi = \varnothing$ and Ψ consisting of a single formula φ.

By a *refutation rule* we mean a rule of the following form.

$$\frac{\vdash \varphi_1, ..., \vdash \varphi_m, \dashv \psi_1, ..., \dashv \psi_n}{\dashv \gamma}$$

Global interpretation (valid/non-valid):
If each φ_i is valid and each ψ_j is non-valid, then γ is non-valid.
A typical example is Łukasiewicz's rule *Reverse modus ponens*:

$$\frac{\vdash \varphi \rightarrow \psi, \dashv \psi}{\dashv \varphi}$$

We now introduce a *local refutation rule* as follows:

$$\frac{\Vdash_1 \varphi_1, \ldots, \Vdash_m \varphi_m, \dashv\!\vdash_1 \psi_1, \ldots, \dashv\!\vdash_n \psi_n}{\dashv\!\vdash_0 \gamma}$$

It has a *local interpretation* (true/false):
Let $\langle \mathcal{F}, V \rangle$ be a finite model. If each φ_i is true at a world $u_i \in W$ and each ψ_j is false at a world $w_j \in W$, then γ is false at the world $w_0 \in W$.

A *refutation system* is a set R of refutation rules, some of them possibly local. Refutation rules with no premises are called *refutation axioms* and we write $\dashv \varphi$, resp. $\dashv\!\vdash_i \varphi$.

We say that a formula φ is R-*refutable* (or just *refutable*) iff there is an R-derivation for $\dashv \varphi$, that is, a sequence $S_1, \ldots, S_t$, where S_t is $\dashv \varphi$ (resp. $\dashv\!\vdash_i \varphi$) and every S_i is either a refutation axiom or has the form $\vdash \vartheta$ (resp. $\Vdash_i \vartheta$) or is obtained from some preceding formulae by a refutation rule.

Let Φ be a set of formulae in some standard logical language L. We say that a refutation system R is *(sound and) complete for* Φ if for any formula φ of the language L the following holds: φ is R-*refutable iff* $\varphi \notin \Phi$.

The refutation system R is complete for a given modal logic **L** if it is complete for the set of valid formulae of **L**. Likewise, for first-order theories.

3 Refutation systems for modal logics in the finite

3.1 Preliminaries: characteristic modal formulae for pointed finite Kripke models

Consider any pointed finite Kripke model $(\mathcal{M}, w)$. For any $n \in \mathbb{N}$, it can be characterised up to n-bisimulation by a single *characteristic modal formula of modal depth* n, denoted $\chi^n_{[\mathcal{M},w]}$ and defined inductively on n as follows (cf. Goranko and Otto (2007)).

The formula $\chi^0_{[\mathcal{M},w]}$ is the conjunction of all $p \in AT$ that are true in w and all $\neg p$ for those that are false at w. It characterises the propositional type of w in $\mathcal{M}$.

Now, suppose the formulae $\chi^n_{[\mathcal{M},u]}$ are defined, for all $u \in \mathcal{M}$. Then we define

$$\chi^{n+1}_{[\mathcal{M},w]} := \chi^0_{[\mathcal{M},w]} \wedge \underbrace{\bigwedge_{(w,u)\in R} \Diamond \chi^n_{[\mathcal{M},u]}}_{forth} \wedge \underbrace{\Box \bigvee_{(w,u)\in R} \chi^n_{[\mathcal{M},u]}}_{back}.$$

Thus, $\chi^{n+1}_{[\mathcal{M},w]}$ describes all successors of w in $\mathcal{M}$ up to depth n.

Let ML^n denote the set of modal formulae of modal depth up to n. Two pointed Kripke models $(\mathcal{M}, w)$ and $(\mathcal{M}', w')$ are called *modally n-equivalent*, denoted $(\mathcal{M}, w) \equiv_{ML^n} (\mathcal{M}', w')$, if they satisfy the same modal formulae of modal dept up to n.

The following claim (cf. Goranko and Otto, 2007) captures the use of the characteristic modal formulae for our purposes.

Proposition 1. *For every two pointed Kripke models $(\mathcal{M}, w)$ and $(\mathcal{M}', w')$ defined on a finite set of atomic propositions the following are equivalent:*

1. $\mathcal{M}', w' \models \chi^n_{[\mathcal{M},w]}$.
2. $(\mathcal{M}, w) \equiv_{\mathsf{ML}^n} (\mathcal{M}', w')$.

3.2 Generic refutation systems for modal logics of finite frames

Consider a finite Kripke frame $\mathcal{F} = \langle W, R \rangle$, where $W = \{w_1, \ldots, w_k\}$.

Denote by $\mathbf{K}_\mathcal{F}$ the normal modal logic of all validities in $\mathcal{F}$.

Let $P = \{p_1, \ldots, p_k\}$ be a set of fixed distinct propositional variables and $\mathcal{F}^P = \langle \mathcal{F}, V^P \rangle$ be the Kripke model where $V^P(p_i) = \{w_i\}$, for $i = 1, \ldots, k$, and $V^P(q) = \varnothing$ for any other propositional variable q. Further, let $\Sigma_\vee(P)$ be the set of all substitutions that replace any propositional variable by a disjunction of variables from P.

Let $\mathsf{Ax}_\mathbf{K}$ be any fixed complete axiomatic system for the basic normal modal logic $\mathbf{K}$.

Theorem 1. *The refutation rule schema*

$$\mathbf{Ref}_\mathcal{F} \quad \frac{\vdash_{\mathsf{Ax}_\mathbf{K}} \chi^n_{[\mathcal{F}^P, w]} \to \neg \sigma(\varphi)}{\dashv \varphi}$$

where $w \in W$, $\varphi \in \mathsf{ML}^n$, and $\sigma \in \Sigma_\vee(P)$, is complete for the modal logic $\mathbf{K}_\mathcal{F}$.

Proof. Suppose φ is semantically refutable in $\mathbf{K}_\mathcal{F}$, i.e. falsifiable at some world w for some valuation V in $\mathcal{F}$. Then there is a substitution $\sigma \in \Sigma_\vee(P)$ such that $\sigma(\varphi)$ fails at w in $\mathcal{F}^P$, viz. the one defined by $\sigma(q) := \bigvee\{p_i \in P \mid w_i \in V(q)\}$ for each $q \in AT$.

Therefore, $\sigma(\varphi)$ fails at every pointed Kripke model $(\mathcal{M}, w)$ defined on P which is n-equivalent to $(\mathcal{F}^P, w)$. By Proposition 1, these are precisely the pointed models $(\mathcal{M}, w)$ defined on P that satisfy $\chi^n_{[\mathcal{F}^P, w]}$. We can assume that $\mathcal{M}$ is extended arbitrarily over AT.

Thus, $\mathcal{M}, w \models \neg \sigma(\varphi)$ for any pointed model $(\mathcal{M}, w)$ such that $M, w \models \chi^n_{[\mathcal{F}^P, w]}$.

Then $\models \chi^n_{[\mathcal{F}^P, w]} \to \neg \sigma(\varphi)$. By completeness of $\mathsf{Ax}_\mathbf{K}$, it follows that $\vdash_\mathbf{K} \chi^n_{[\mathcal{F}^P, w]} \to \neg \sigma(\varphi)$.

Therefore, by applying the rule $\mathbf{Ref}_\mathcal{F}$ we derive $\dashv \varphi$. □

Theorem 1 provides a generic, simple and complete refutation system for every modal logic $\mathbf{K}_\mathcal{F}$, which only employs derivations in $\mathsf{Ax}_\mathbf{K}$. However, it has some clear shortcomings, which are the price to pay for the generality:

- Involves infinitely many (one for each n) and rather complex refutation rules. Moreover, they have no subformula property.

- Does not build the refutable formulae step-by-step, but derives them at once.
- Most essential: to derive ⊣ φ it requires guessing the falsifying substitution σ, which amounts to finding a refuting valuation for φ in $\mathcal{F}$.

3.3 An alternative construction of a refutation system for the modal logic of a finite frame

Again, consider a finite Kripke frame $\mathcal{F} = \langle W, R \rangle$, where $W = \{w_1, \ldots, w_k\}$. We will use the notation from Section 3.2. and will now build in a uniform way a different refutation system $\mathbf{Ref}(\mathcal{F})$ for $\mathcal{F}$, by adding 'local' rules for every state of $\mathcal{F}$, in order to generate all formulae falsifiable at that state from the formulae falsifiable at its successors in the frame.

The refutation system $\mathbf{Ref}(\mathcal{F})$ employs local refutation rules, using an additional auxiliary symbol ⊣| which will be used in combination with any substitution $\sigma \in \Sigma_V(P)$. Intuitively, w⊣|$_\sigma \varphi$ means that φ is falsified at the state w by a valuation syntactically described by the substitution σ.

The system $\mathbf{Ref}(\mathcal{F})$ and consists of several schemes of refutation rules listed below, where:

- ⊢$_{\mathsf{PL}}$ is any complete deduction system for PL.
- ⊢$_{\Box\mathsf{PL}}$ refers to derivability in ⊢$_{\mathsf{PL}}$, but applied to modal formulae of ML, where all subformulae of the type $\Box \psi$ are treated as atomic symbols (recall that $\Diamond$ is assumed defined as $\neg \Box \neg$).
- σ is any substitution in $\Sigma_V(P)$.

The refutation rules of $\mathbf{Ref}(\mathcal{F})$:

1. For every $w_i, w \in W$ and a classical propositional formula β,

$$\mathbf{Ref}^0(w_i) \quad \frac{\vdash_{\mathsf{PL}} p_i \to \sigma(\neg \beta)}{w_i \dashv|_\sigma \beta}$$

and

$$\mathbf{Ref}^1(w) \quad \frac{w \dashv|_\sigma \psi, \ \vdash_{\Box\mathsf{PL}} \varphi \to \psi}{w \dashv|_\sigma \varphi}$$

2. For every $w, w' \in W$ such that wRw':

$$\mathbf{Ref}^\Box(w, w') \quad \frac{w \dashv|_\sigma \varphi, \ w' \dashv|_\sigma \psi}{w \dashv|_\sigma \varphi \vee \Box \psi}$$

3. For every $w \in W$ and $\{u_1, \ldots, u_m\}$ as its set of all R-successors of w:

$$\mathbf{Ref}^\Diamond(w) \quad \frac{w \dashv|_\sigma \varphi, \ u_1 \dashv|_\sigma \psi, \ldots, u_m \dashv|_\sigma \psi}{w \dashv|_\sigma \varphi \vee \Diamond \psi}$$

4. Finally, for every $w \in W$:

$$\mathbf{Ref}(w) \quad \frac{w\dashv\!\!\|_\sigma \varphi}{\dashv \varphi}$$

Theorem 2. *The refutation system* $\mathbf{Ref}(\mathcal{F})$ *is complete for the logic* $\mathbf{K}_\mathcal{F}$.

Proof. First, we note, by routine induction on derivations in $\mathbf{Ref}(\mathcal{F})$, that whenever $w\dashv\!\!\|_\sigma \varphi$ is derivable then φ is falsifiable in $\mathcal{F}$ at the state w by the valuation V_σ defined by $V_\sigma(q) = \{w_i \in W \mid p_i \text{ occurs as a disjunct in } \sigma(q)\}$. Therefore, if $\dashv \varphi$ then φ is falsifiable in $\mathcal{F}$ at w by the valuation V_σ, hence φ is semantically refutable in $\mathbf{K}_\mathcal{F}$.

For the converse, first we define for any valuation V in $\mathcal{F}$ the substitution $\sigma_V \in \Sigma_V(P)$ as in the proof of Thm 1, viz. $\sigma_V(q) := \bigvee\{p_i \in P \mid w_i \in V(q)\}$ for each $q \in AT$. Note that for every modal formula φ and $w \in W$:
$\langle \mathcal{F}, V \rangle, w \models \varphi$ iff $\langle \mathcal{F}, V^P \rangle, w \models \sigma_V(\varphi)$.

Now, we will prove by induction on $d \in \mathbb{N}$ that for every $\varphi \in \mathrm{ML}^d$, if φ is falsifiable in $\mathcal{F}$ at some state w by a valuation V then $w\dashv\!\!\|_{\sigma_V}\varphi$ is derivable in $\mathbf{Ref}(\mathcal{F})$, hence $\dashv \varphi$ is derivable in $\mathbf{Ref}(\mathcal{F})$, by $\mathbf{Ref}(w)$.

First, let $d = 0$ and $\varphi \in \mathrm{ML}^0$ be a propositional formula falsifiable by V at some $w_i \in W$. Then $\langle \mathcal{F}, V^P \rangle, w_i \models \sigma_V(\neg \varphi)$, hence $\models p_i \to \sigma_V(\neg \varphi)$, so $\vdash_{\mathsf{PL}} p_i \to \sigma_V(\neg \varphi)$. Therefore $w_i \dashv\!\!\|_{\sigma_V}\varphi$ by $\mathbf{Ref}^0(w)$.

Suppose the claim holds for all formulae in ML^d for some $d \in \mathbb{N}$.

Now, let $\varphi \in \mathrm{ML}^{d+1}$ be falsifiable at some state $w \in \mathcal{F}$ by some valuation V, i.e., $\langle \mathcal{F}, V \rangle, w \models \neg \varphi$. Note that φ is propositionally equivalent to a conjunction of formulae of the type

$$\theta = \beta \vee \Diamond\varphi_1 \vee \ldots \vee \Diamond\varphi_m \vee \Box\psi_1 \vee \ldots \vee \Box\psi_n,$$

where $m, n \geq 0$ and $\varphi_1, \ldots, \varphi_m, \psi_1, \ldots, \psi_n \in \mathrm{ML}^d$.

Then, at least one of these conjuncts θ is false at w in $(\mathcal{F}, V)$. We fix it.

We will show that $w\dashv\!\!\|_{\sigma_V}\theta$ is derivable in $\mathbf{Ref}(\mathcal{F})$.

To begin with, we have $\langle \mathcal{F}, V^P \rangle, w \models \sigma_V(\neg \beta)$, hence $w\dashv\!\!\|_{\sigma_V}\beta$ is derivable in $\mathbf{Ref}(\mathcal{F})$, by using $\mathbf{Ref}^0(w)$ as in the case $d = 0$.

Next, each ψ_i is falsified in $\langle \mathcal{F}, V \rangle$ at some successor w^i of w and each φ_j is falsified in $\langle \mathcal{F}, V \rangle$ at all successors u of w. By the inductive hypothesis, it follows that $w^i\dashv\!\!\|_{\sigma_V}\psi_i$, for each $i = 1, \ldots, n$, and $u\dashv\!\!\|_{\sigma_V}\varphi_j$ for each $j = 1, \ldots, m$ and each successor u of w, are all derivable in $\mathbf{Ref}(\mathcal{F})$.

Now, by applying $\mathbf{Ref}^\Diamond(w)$ consecutively, for $\psi = \varphi_1, \ldots, \varphi_m$, we derive $w\dashv\!\!\|_{\sigma_V}\beta \vee \Diamond\varphi_1 \vee \ldots \vee \Diamond\varphi_m$.

Then, likewise, by applying $\mathbf{Ref}^\Box(w, w')$ consecutively, for $w' = w^1, \ldots, w^m$, we eventually derive $w\dashv\!\!\|_{\sigma_V}\theta$ in $\mathbf{Ref}(\mathcal{F})$.

Finally, note that $\vdash_{\Box\mathsf{PL}} \varphi \to \theta$, hence $w\dashv\!\!\|_{\sigma_V}\varphi$ is derivable in $\mathbf{Ref}(\mathcal{F})$, by $\mathbf{Ref}^1(w)$, which completes the induction, and the entire proof. □

3.4 On refutation systems for modal logics with finite model property

Each of the refutation systems described in sections 3.2 and 3.3 extends to the modal logic **K** of any recursively enumerable set of finite Kripke frames, by adding the respective refutation rules for each frame $\mathcal{F} \in$.

In particular, using this generic construction, a complete generic refutation system can likewise be produced for any (finitely axiomatized) normal modal logic with finite frame property[1]. Interestingly, even finite axiomatisation is not needed (and, possibly, not even existing) when the class of all finite frames of the logic is explicitly known. An important case in point is Medvedev's logic (see Skura, 1992), the decidability of which is not known yet, but, being defined on a class of simply and explicitly defined finite frames, it has a recursively enumerable set of non-valid formulae. Actually, the following simple refutation system is complete for this set (see Skura, 1992).

Refutation axiom: $\bot$ (the false).

Refutation rules:

- **(RS)** (If a substitution instance of φ is refutable, so is φ.)
- **(RMP$_{KP}$)** $\vdash_{KP} \varphi \to \psi, \dashv \psi / \dashv \varphi$
  (Here $\vdash_{KP}$ means provability in the Kreisel-Putnam logic.)
- **(RD)** $\dashv \varphi, \dashv \psi / \dashv \varphi \vee \psi$

4 Refutation system for FO in the finite

4.1 Refutation system for the FO theory of a single finite model

Fix any FO language L with finite signature.

Definition 1. Let $n \in \mathbb{N}^+$. For any list of distinct variables $\bar{x} = \{x_1, ..., x_n\}$ we define:

1. $\Gamma_n(\bar{x}) = \{x_i = x_j : 1 \leq i < j \leq n\}$,
2. $\delta_n(\bar{x}) = \forall x_{n+1}(x_{n+1} = x_1 \vee ... \vee x_{n+1} = x_n)$.
3. $\Theta_n(\bar{x})$ is the set of **basic atomic formulae of** L **over** $\bar{x}$, consisting of all formulae of the form:

   a. $x_i = x_j$, or
   b. $c = x_j$ for any constant symbol $c \in L$, or
   c. $R y_1, ..., y_k$ for any k-ary relational symbol $R \in L$, or
   d. $f(y_1, ..., y_k) = y$ for any k-ary functional symbol $f \in L$,
   with variables $y_1, ..., y_k, y \in \{x_1, ..., x_n\}$.

---

[1] The finite axiomatization is only needed in order to be able to recognise the finite frames of that logic.

Definition 2. Consider a finite L-structure $\mathcal{A}$ with universe $A = \{a_1, ..., a_n\}$. We define the FO sentence $\chi_{\mathcal{A}}$ that describes $\mathcal{A}$ up to isomorphism (see e.g. (Ebbinghaus and Flum, 2005, p. 13), where it is denoted $\varphi_{\mathcal{A}}$), as follows:

$$\chi_{\mathcal{A}} := \exists x_1 ... \exists x_n \left(\delta_n(\bar{x}) \wedge \bigwedge \{\psi \mid \psi \in \mathsf{POS}(\mathcal{A})\} \wedge \bigwedge \{\neg \psi \mid \psi \in \mathsf{NEG}(\mathcal{A})\} \right),$$

where $\bar{x} = \{x_1, ..., x_n\}$ is a list of distinct variables,
$\bar{x} := \bar{a}$ is the assignment α in $\mathcal{A}$ such that $\alpha(x_1) = a_1, ..., \alpha(x_n) = a_n$,
$\mathsf{POS}(\mathcal{A}) = \{\psi \mid \psi \in \Theta_n(\bar{x}) \text{ and } \mathcal{A} \models \psi[\bar{x} := \bar{a}]\}$, and
$\mathsf{NEG}(\mathcal{A}) = \{\psi \mid \psi \in \Theta_n(\bar{x}) \text{ and } \mathcal{A} \models \neg\psi[\bar{x} := \bar{a}]\}$

Note that every formula $x_i = x_j$, where $i \neq j$, is in $\mathsf{NEG}(\mathcal{A})$.

Hereafter we fix any complete deductive system $\mathsf{Ax_{FO}}$ for FO.

Lemma 1. *Let $\mathcal{A}$ be a finite structure and φ be a sentence which is not true in $\mathcal{A}$. Then*

$$\vdash_{\mathsf{Ax_{FO}}} \varphi \rightarrow \neg \chi_{\mathcal{A}}$$

Proof. The sentence $\neg \varphi$ is true in every FO structure isomorphic to $\mathcal{A}$, i.e. satisfying $\chi_{\mathcal{A}}$. Therefore, $\models \chi_{\mathcal{A}} \rightarrow \neg \varphi$, hence $\models \varphi \rightarrow \neg \chi_{\mathcal{A}}$. Then the claim follows by completeness of $\mathsf{Ax_{FO}}$. □

Theorem 3. *Let $\mathcal{A}$ be any finite FO structure. Then the refutation system consisting of*

$$\mathbf{Ref}_{\mathcal{A}} \quad \dfrac{\vdash_{\mathsf{Ax_{FO}}} \chi_{\mathcal{A}} \rightarrow \neg \varphi}{\dashv \varphi}$$

and **Reverse generalisation**:

$$\mathbf{RG} \quad \dfrac{\dashv \forall x \varphi}{\dashv \varphi}$$

is complete for the FO theory TH($\mathcal{A}$) of $\mathcal{A}$.

Proof. First, let φ be a sentence which is not true in $\mathcal{A}$. Then $\vdash_{\mathsf{Ax_{FO}}} \chi_{\mathcal{A}} \rightarrow \neg \varphi$, by Lemma 1. Then, by applying $\mathbf{Ref}_{\mathcal{A}}$, we obtain $\dashv \varphi$.

Now, let $\varphi = \varphi(\bar{x})$ be any formula with a tuple of free variables $\bar{x} = \langle x_1, ..., x_k \rangle$, which is not true in $\mathcal{A}$ and let $\forall \bar{x} \varphi$ be a universal closure of φ. Then, $\mathcal{A} \not\models \forall \bar{x} \varphi$. Hence, by the argument above, $\dashv \forall \bar{x} \varphi$. By using the rule $\mathbf{RG}$ repeatedly, we eventually obtain $\dashv \varphi$. □

Just like with modal logic, the refutation system above extends readily to the FO theory of any recursively enumerable set of finite FO structures, by adding refutation rules $\mathbf{Ref}_{\mathcal{A}}$ for each $\mathcal{A} \in$. In particular, a complete refutation system can thus be produced for all FO formulae that are not valid in the finite. That system, however, involves infinitely many refutation rules $\mathbf{Ref}_{\mathcal{A}}$ and is not practically very useful. In the next subsection we will develop an alternative, purely syntactic refutation system for the FO on all finite structures.

4.2 Refutation system $\mathbf{Ref}_{FO}^{fin}$ for FO in the finite

Let FO^{fin} be the set of all first-order formulae true in all finite structures under all assignments, and let RFO^{fin} be its complement, i.e. the set of first-order formulae falsifiable in some finite structure under some assignment.

By Trakhtenbrot's theorem, FO^{fin} is not recursively enumerable, hence it cannot have a complete recursive axiom system. However, RFO^{fin} is recursively enumerable and therefore it can be recursively axiomatized. In this section we give a refutation system for the set of the formulae of FO logic that are not valid in the finite.

First, some preliminaries. Hereafter, $\vdash$ stands for provability in some fixed complete deductive system Ax_{FO} for FO.

We now introduce the following refutation system Ref_{FO}^{fin} for FO.

Refutation axioms. (Recall $\Gamma_n(\bar{x})$ and $\Theta_n(\bar{x})$ from Definition 1.)

$$\dashv \Phi, \delta_n(\bar{x}) \longrightarrow \Gamma_n(\bar{x}) \cup \Psi$$

where $n \geq 1$ and $\Phi \cup \Psi$ is a finite set of basic atomic formulae from $\Theta_n(\bar{x})$, such that:

1. $\Phi \cap \Psi = \emptyset$.
2. $\Phi \cup \Psi$ does not contain formulae of the type $x_i = x_j$.
3. For every constant symbol $c \in L$, exactly one formula of the type $c = x_i$ is in Φ.
4. For every k-ary functional symbol $f \in L$ and $y_1, ..., y_k \in \{x_1, ..., x_n\}$, exactly one formula of the type $f(y_1, ..., y_k) = x_i$ is in Φ.

Refutation rules:

Reverse modus ponens (**RMP**):

$$\frac{\vdash \varphi \to \psi \quad \dashv \psi}{\dashv \varphi}$$

Reverse generalisation (**RG**):

$$\frac{\dashv \forall x \varphi(x)}{\dashv \varphi(x)}$$

Lemma 2. *If φ is a refutation axiom, then $\varphi \in RFO^{fin}$.*

Proof. Let $\varphi = \Phi, \delta_n(\bar{x}) \longrightarrow \Gamma_n(\bar{x}) \cup \Psi$ be a refutation axiom over $x_1, ..., x_n$. Consider the n-element structure $\mathcal{A}$ with domain $A = \{a_1, ..., a_n\}$, where:

1. For every constant symbol $c \in L$, its interpretation $c^{\mathcal{A}}$ is the (unique) $a_i \in A$ such that $c = x_i$ is in Φ.
2. For every k-ary functional symbol $f \in L$ its interpretation $f^{\mathcal{A}}$ is defined as follows: for every tuple $\langle a_{i_1}, ..., a_{i_k} \rangle \in A^k$, $f^{\mathcal{A}}(a_{i_1}, ..., a_{i_k})$ is the (unique) $a_i \in A$ such that $f(x_{i_1}, ..., x_{i_k}) = x_i$ is in Φ.

3. For every k-ary relational symbol $R \in L$ its interpretation $R^\mathcal{A}$ is defined as follows: for every tuple $\langle a_{i_1}, ..., a_{i_k} \rangle \in A^k$, $R^\mathcal{A}(a_{i_1}, ..., a_{i_k})$ holds iff $Rx_{i_1}, ..., x_{i_k} \in \Phi$.

Consider the assignment α in $\mathcal{A}$, where $\alpha(x_i) := a_i$, for $i = 1, ..., n$. Then:

- $\mathcal{A} \models \delta_n(\bar{x})[\alpha]$.
- If $\theta \in \Phi$ then $\mathcal{A} \models \theta[\alpha]$.
- If $\theta \in \Gamma_n(\bar{x}) \cup \Psi$ then $\mathcal{A} \not\models \theta[\alpha]$.

Hence, $\mathcal{A} \not\models \varphi[\alpha]$. □

Lemma 3. *Let φ be any formula. Then $\forall x_i \varphi \in \mathsf{RFO}^{\mathsf{fin}}$ iff $\varphi \in \mathsf{RFO}^{\mathsf{fin}}$.*

Proof. Straightforward. □

Theorem 4. *Let φ be any formula. Then $\varphi \in \mathsf{RFO}^{\mathsf{fin}}$ iff $\dashv \varphi$.*

Proof. ($\Leftarrow$) holds since the refutation axioms are not in $\mathsf{FO}^{\mathsf{fin}}$, and the set $\mathsf{RFO}^{\mathsf{fin}}$ is closed under the refutation rules, by Lemma 3.

($\Rightarrow$) Let $\varphi \in \mathsf{RFO}^{\mathsf{fin}}$. First, suppose φ is a sentence. Then φ is falsified by some finite structure $\mathcal{A}$, with universe $\{a_1, ..., a_n\}$.

By Lemma 1, we have $\vdash \varphi \rightarrow \neg \chi_\mathcal{A}$. (1)

Note that $\vdash \neg \chi_\mathcal{A} \leftrightarrow \forall x_1 ... \forall x_n \mathsf{NDIAG}(\mathcal{A})$, where (recall Def. 2)

$$\mathsf{NDIAG}(\mathcal{A}) = \mathsf{POS}(\mathcal{A}), \delta_n(\bar{x}) \longrightarrow \mathsf{NEG}(\mathcal{A})$$

Hence, $\vdash \neg \chi_\mathcal{A} \rightarrow \forall x_1 ... \forall x_n \mathsf{NDIAG}(\mathcal{A})$. (2)

Note that $\mathsf{NDIAG}(\mathcal{A})$ is a refutation axiom. Thus, $\dashv \mathsf{NDIAG}(\mathcal{A})$.
Hence, $\dashv \forall x_1 ... \forall x_n \mathsf{NDIAG}(\mathcal{A})$, by repeated application of **RMP**, using $\vdash \forall x \varphi \rightarrow \varphi$.

Then, $\dashv \neg \chi_\mathcal{A}$, by (2) and **RMP**.
Therefore, $\dashv \varphi$, by (1) and **RMP**.

Now, let $\varphi \in \mathsf{RFO}^{\mathsf{fin}}$, where φ is any formula, and let $\forall \bar{x} \varphi$ be a universal closure of φ. Then, $\forall \bar{x} \varphi \in \mathsf{RFO}^{\mathsf{fin}}$. Hence, by the argument above, $\dashv \forall \bar{x} \varphi$. By using the rule **RG** repeatedly, we eventually obtain $\dashv \varphi$. □

Acknowledgments

This work was partly supported by research grant 2015-04388 of the Swedish Research Council. We thank the anonymous referee for some corrections.

References

Barwise J (1977) An introduction to first-order logic. In: Studies in Logic and the Foundations of Mathematics, vol 90, Elsevier, pp 5–46

Citkin A (2013) Jankov-style formulas and refutation systems. Reports on Mathematical Logic (48):67–80

Ebbinghaus HD, Flum J (2005) Finite model theory. Springer Science & Business Media

Goranko V (1994) Refutation systems in modal logic. Studia Logica 53(2):299–324

Goranko V (2016) Logic as a Tool: A Guide to Formal Logical Reasoning. John Wiley & Sons

Goranko V, Otto M (2007) 5 model theory of modal logic. In: Studies in Logic and Practical Reasoning, vol 3, Elsevier, pp 249–329

Goré R, Postniece L (2008) Combining derivations and refutations for cut-free completeness in bi-intuitionistic logic. Journal of Logic and Computation 20(1):233–260

Hailperin T (1961) A complete set of axioms for logical formulas invalid in some finite domain. Mathematical Logic Quarterly 7(6):84–96

Skura T (2011) Refutation systems in propositional logic. In: Handbook of Philosophical Logic, Springer, pp 115–157

Skura T (2013) Refutation methods in modal propositional logic. Semper Warsaw

Skura T (1992) Refutation calculi for certain intermediate propositional logics. Notre Dame Journal of Formal Logic 33(4):552–560

Skura T (1994) Syntactic refutations against finite models in modal logic. Notre Dame Journal of Formal Logic 35(4):595–605

Tiomkin M (1988) Proving unprovability. In: [1988] Proceedings. Third Annual Symposium on Logic in Computer Science, IEEE, pp 22–26

Towards a Uniform Account of Proofs and Refutations

Andrzej Wiśniewski

Abstract A set of well-formed formulas (wffs) is holistically inconsistent iff it is inconsistent, but each wff in the set is consistent. We present a sequent calculus for holistically inconsistent sets of wffs of Classical Propositional Logic. Since valid, inconsistent, and contingent wffs correspond to different, yet strictly defined, holistically inconsistent sets, a proof of a sequent based on holistically inconsistent set of a given kind can be regarded, depending on the case, as a proof of a valid wff, a refutation of an inconsistent wff, and a refutation of a contingent wff, respectively.

Key words: proofs, refutations, holistic inconsistency, sequent calculi

1 Introduction

Consider a formal language supplemented with a bivalent semantics rich enough to define some concept of truth of a well-formed formula (henceforth: wff) in a model. The expression "model" is used here as a cover term; depending on the particular form of the language, models are valuations of some kind, relational structures, and so on. Usually, a formal language has many models of a given kind. When a non-empty class of models, $\mathcal{M}$, is fixed, the set of all wffs of the language splits, first, into two disjoint subsets: $\mathtt{Val}^{\mathcal{M}}$ and $\mathtt{NVal}^{\mathcal{M}}$. The set $\mathtt{Val}^{\mathcal{M}}$ comprises all the wffs which are *valid* w.r.t. the class of models $\mathcal{M}$, that is, which are true in each model from $\mathcal{M}$. The set $\mathtt{NVal}^{\mathcal{M}}$, in turn, comprises all the remaining wffs, that is, wffs which are not valid w.r.t. the class of models $\mathcal{M}$. However, the set $\mathtt{NVal}^{\mathcal{M}}$ is far from being homogenous. It includes *inconsistent* (also called *unsatisfiable*) wffs, that is, wffs which are not true

---

Andrzej Wiśniewski
Department of Logic and Cognitive Science
Faculty of Psychology and Cognitive Science
Adam Mickiewicz University, Poznań, Poland
e-mail: Andrzej.Wisniewski@amu.edu.pl

in any model from the class $\mathcal{M}$. But it also includes wffs which are consistent (or *satisfiable*) without being valid, i.e. wffs which are true in some model(s) belonging to the class $\mathcal{M}$, but are not true in other models from the class. Following a philosophical rather than a logical tradition, let us call these wffs *contingent*. To be more precise, when a class of models $\mathcal{M}$ is fixed, the set $\mathrm{NVal}^{\mathcal{M}}$ splits into the set $\mathrm{Inc}^{\mathcal{M}}$ of $\mathcal{M}$-*inconsistent* wffs (i.e. wffs which are not true in any model from $\mathcal{M}$) and the set $\mathrm{Ctg}^{\mathcal{M}}$ of $\mathcal{M}$-*contingent* wffs, that is, wffs which are neither $\mathcal{M}$-valid nor $\mathcal{M}$-inconsistent.

Looking from the proof-theoretic point of view, the main challenge for a logician is to build a calculus which makes provable all the valid (w.r.t. a given class of models) wffs and only them. Sometimes, as a by-product, a calculus gives an account of inconsistent wffs as well. Analytic tableaux are paradigmatic examples here. However, contingent wffs remain beyond the scope of interest. The (still rare) advocates of refutation methods see the goal differently: they aim at proof-theoretic accounts of non-validities (cf. Skura (2009), Skura (2011)). But the class of non-validities includes both inconsistent wffs and contingent wffs. This distinction seems to play no role in refutation calculi, however. Last but not least, logical calculi focussed on validities and these focussed on non-validities operate with diverse formal means.

The aim of this short note is to present a calculus which, on the one hand, differentiates between proofs of valid wffs, refutations of inconsistent wffs, and refutations of contingent wffs. On the other hand, the calculus offers a uniform proof-mechanism. This is achieved by the introduction of a kind of conceptual unifier, namely the notion of a holistically inconsistent *set* of wffs. The system "calculates" such sets or, more precisely, sequents based on them. Since valid, inconsistent, and contingent wffs correspond to different, yet strictly defined, holistically inconsistent sets, a proof of a sequent based on a set of a given kind can be regarded, depending on the case, as a proof or as a refutation of the corresponding wff.

2 The logical basis

We remain at the level of Classical Propositional Calculus (CPL for short). As for the language of (the analysed version of) CPL, we assume that the vocabulary comprises a countably infinite set of propositional variables, the connectives: $\neg, \vee, \wedge, \rightarrow, \leftrightarrow$, and brackets. *Well-formed formulas* (henceforth: wffs) of the language are defined as usual. We use A, B, C, D, with subscripts when needed, as metalanguage variables for wffs, and X, Y, with or without subscripts or superscripts, as metalanguage variables for sets of wffs. The letters p, q, r, s, t are exemplary elements of the set of propositional variables of the language.

Let **1** stand for truth and **0** for falsity. A CPL-*valuation* is a function from the set of wffs to the set $\{\mathbf{1}, \mathbf{0}\}$, satisfying the following standard conditions: (a) $v(\neg A) = \mathbf{1}$ iff $v(A) = \mathbf{0}$; (b) $v(A \vee B) = \mathbf{1}$ iff $v(A) = \mathbf{1}$ or $v(B) = \mathbf{1}$; (c) $v(A \wedge B) = \mathbf{1}$ iff $v(A) = \mathbf{1}$ and $v(B) = \mathbf{1}$; (d) $v(A \rightarrow B) = \mathbf{1}$ iff $v(A) = \mathbf{0}$ or $v(B) = \mathbf{1}$, (e) $v(A \leftrightarrow B) = \mathbf{1}$ iff $v(A) = v(B)$.

Towards a Uniform Account of Proofs and Refutations   183

For brevity, in what follows we will be omitting references to CPL. By wffs we will mean wffs of the language of CPL, and by valuations we will mean CPL-valuations.

Definition 1 (Consistency, inconsistency, validity, and contingence). A set of wffs X is consistent iff there exists a valuation v such that for each $A \in X$, $v(A) = 1$; otherwise X is inconsistent. A wff B is:

1. consistent iff the singleton set $\{B\}$ is consistent,
2. inconsistent iff the singleton set $\{B\}$ is inconsistent,
3. valid iff for each valuation v, $v(B) = 1$,
4. contingent iff B is neither inconsistent nor valid.

CPL-entailment, $\models$, is defined as follows:

Definition 2 (Entailment). $X \models A$ iff for each valuation v:

- if $v(B) = 1$ for every $B \in X$, then $v(A) = 1$.

The next definition introduces the crucial notion.

Definition 3 (Holistically inconsistent set; HI-set). A set of wffs X is holistically inconsistent iff X is inconsistent, but each wff in X is consistent.

Observe that each HI-set has at least two elements.

The following are true:

Corollary 1. *A wff C is contingent iff $\{C, \neg C\}$ is a HI-set.*

Proof. ($\Rightarrow$) If C is a contingent wff, then there are valuations v, v^* such that $v(C) = 1$ and $v^*(C) = 0$. So both C and $\neg C$ are consistent wffs. On the other hand, the set $\{C, \neg C\}$ is inconsistent. Therefore $\{C, \neg C\}$ is a HI-set.
($\Leftarrow$) If $\{C, \neg C\}$ is a HI-set, then both C and $\neg C$ are consistent wffs. Thus C is a contingent wff. □

Corollary 2. *A wff C is inconsistent iff $\{C \vee p, C \vee \neg p\}$ is a HI-set.*

Proof. ($\Rightarrow$) Assume that C is an inconsistent wff. Each of the wffs: $C \vee p$, $C \vee \neg p$, is consistent, however. On the other hand, the set $\{C \vee p, C \vee \neg p\}$ is inconsistent and hence is a HI-set.
($\Leftarrow$) If $\{C \vee p, C \vee \neg p\}$ is a HI-set, it is an inconsistent set and hence $\{C \vee p, C \vee \neg p\} \models p \wedge \neg p$. It follows that $\{C, C \vee \neg p\} \models p \wedge \neg p$ and therefore $C \models p \wedge \neg p$. Thus C is inconsistent. □

Corollary 3. *A wff C is valid iff $\{\neg C \vee p, \neg C \vee \neg p\}$ is a HI-set.*

Proof. ($\Rightarrow$) If C is valid, then $\neg C$ is inconsistent. But, similarly as before, both $\neg C \vee p$ and $\neg C \vee \neg p$ are consistent wffs, and the set $\{\neg C \vee p, \neg C \vee \neg p\}$ is inconsistent. Therefore $\{\neg C \vee p, \neg C \vee \neg p\}$ is a HI-set.
($\Leftarrow$) The set $\{\neg C \vee p, \neg C \vee \neg p\}$, as a HI-set, is inconsistent. Thus $\{\neg C \vee p, \neg C \vee \neg p\} \models p \wedge \neg p$ and therefore $\neg C \models p \wedge \neg p$. It follows that $\neg C$ is an inconsistent wff and hence C is a valid wff. □

Thus validity, inconsistency and contingency of wffs are expressible in terms of HI-sets.

3 The system HI^{CPL}

Since the system we are going to present "calculates" HI-sets, we label it by HI^{CPL}.

We operate with *sequents* of the form $Y \vdash$, where Y is an at least two-element finite set of CPL-wffs. In practice, we write down a sequent $Y \vdash$ by listing the elements of Y left to the turnstile. An inscription of the form '$C \in$ CPL' means 'C is a thesis of CPL', i.e. is provable in CPL.

A sequent, $Y \vdash$, is in the *normal form* iff each $C \in Y$ is in the disjunctive normal form (hereafter: DNF).

An *axiom* of HI^{CPL} is a sequent $Y \vdash$ such that each $B \in Y$ is an elementary conjunction, a conjunction of all the wffs in Y involves complementary literals, and no $B \in Y$ involves complementary literals. Here are examples of axioms:

$$p, \neg p \vdash \tag{1}$$

$$\neg p \wedge \neg q, p \vdash \tag{2}$$

$$\neg p \wedge \neg q, q \vdash \tag{3}$$

$$\neg p \wedge \neg q, p \wedge \neg q, q \wedge \neg p \vdash \tag{4}$$

There are only two (primary) *rules* of HI^{CPL}, namely:

R$_1$: $\dfrac{Y \cup \{A\} \vdash \quad Y \cup \{B\} \vdash}{Y \cup \{A \vee B\} \vdash}$

R$_2$: $\dfrac{Y \cup \{A\} \vdash}{Y \cup \{B\} \vdash}$ where $(A \leftrightarrow B) \in$ CPL.

Definition 4 (Proof of a sequent). A proof of a sequent $Y \vdash$ in HI^{CPL} is a finite labelled tree regulated by the rules of HI^{CPL}, where the leaves are labelled with axioms and $Y \vdash$ labels the root.

A sequent $Y \vdash$ is provable in HI^{CPL} iff the sequent $Y \vdash$ has at least one proof in HI^{CPL}.

Here are examples of proofs:

Example 1. A proof of the sequent $\neg(p \vee q), \neg(\neg p \wedge \neg q) \vdash$:

$$\neg p \wedge \neg q, p \vdash \qquad \neg p \wedge \neg q, q \vdash$$
$$\neg p \wedge \neg q, \neg\neg p \vdash \qquad \neg p \wedge \neg q, \neg\neg q \vdash$$
$$\neg p \wedge \neg q, \neg\neg p \vee \neg\neg q \vdash$$
$$\neg p \wedge \neg q . \neg(\neg p \wedge \neg q) \vdash$$
$$\neg(p \vee q), \neg(\neg p \wedge \neg q) \vdash$$

Example 2. A proof of the sequent $(p \lor q) \land \neg p, (p \lor q) \land \neg q, \neg p \land \neg q \vdash$:

$$q \land \neg p, p \land \neg q, \neg p \land \neg q \vdash$$
$$q \land \neg p, (q \land \neg q) \lor (p \land \neg q), \neg p \land \neg q \vdash$$
$$q \land \neg p, (p \lor q) \land \neg q, \neg p \land \neg q \vdash$$
$$(q \land \neg p) \lor (p \land \neg p), (p \lor q) \land \neg q, \neg p \land \neg q \vdash$$
$$(p \lor q) \land \neg p, (p \lor q) \land \neg q, \neg p \land \neg q \vdash$$

Provability of a sequent $Y \vdash$ yields that Y is HI-set. This is due to

Theorem 1 (Soundness of HI^{CPL} w.r.t. HI-sets). *Let Y be an at least two element finite set of wffs. If the sequent $Y \vdash$ is provable in HI^{CPL}, then Y is a HI-set.*

Proof. Clearly, if $Y \vdash$ is an axiom, then Y is a HI-set.

Assume that $Y \cup \{A\}$ and $Y \cup \{B\}$ are HI-sets. Thus each wff in Y is consistent. Moreover, the set $Y \cup \{A \lor B\}$ is inconsistent – otherwise $Y \cup \{A\}$ would be consistent or $Y \cup \{B\}$ would be consistent. Suppose that the set $Y \cup \{A \lor B\}$ contains an inconsistent wff. Since each wff in Y is consistent, it follows that $A \lor B$ is inconsistent, and hence both A and B are inconsistent. But in this case neither $Y \cup \{A\}$ nor $Y \cup \{B\}$ is a HI-set. A contradiction.

It is obvious that if $X \cup \{A\}$ is a HI-set and $(A \leftrightarrow B) \in$ CPL, then $X \cup \{B\}$ is a HI-set. □

Theorem 1 together with corollaries 3, 2 and 1 yield:

Theorem 2.

1. *If the sequent $\neg C \lor p, \neg C \lor \neg p \vdash$ is provable in HI^{CPL}, then C is a valid wff.*
2. *If the sequent $C \lor p, C \lor \neg p \vdash$ is provable in HI^{CPL}, then C is an inconsistent wff.*
3. *If the sequent $C, \neg C \vdash$ is provable in HI^{CPL}, then C is a contingent wff.*

The next step is a non-standard one. We define provability of a wff in terms of provability of a sequent of a strictly defined form. But, contrary to what is usually done, we *do not* construe the provability of a wff C as the provability of the sequent based on C or the negation of C only. The definition runs as follows:

Definition 5 (Proof of a wff). A HI^{CPL}-proof of a wff C is a proof of the sequent $\neg C \lor p, \neg C \lor \neg p \vdash$ in HI^{CPL}.

Example 3. A proof of $p \to p$:

$$p, \neg p \vdash$$
$$(p \land \neg p) \lor p, \neg p \vdash$$
$$(p \land \neg p) \lor p, (p \land \neg p) \lor \neg p \vdash$$
$$\neg(p \to p) \lor p, (p \land \neg p) \lor \neg p \vdash$$
$$\neg(p \to p) \lor p, \neg(p \to p) \lor \neg p \vdash$$

Similarly, we define refutability in terms of provability of sequents of strictly defined form. This time, however, we introduce two concepts.

Definition 6 (Refutation[1] of a wff). A $\mathsf{HI}^{\mathsf{CPL}}$-refutation[1] of a wff C is a proof of the sequent $C \vee p, C \vee \neg p \vdash$ in $\mathsf{HI}^{\mathsf{CPL}}$.

Definition 7 (Refutation[2] of a wff). A $\mathsf{HI}^{\mathsf{CPL}}$-refutation[2] of a wff C is a proof of the sequent $C, \neg C \vdash$ in $\mathsf{HI}^{\mathsf{CPL}}$.

Example 4. A refutation[1] of $\neg(p \to p)$:

$$p, \neg p \vdash$$
$$\neg(\neg p \vee p) \vee p, \neg p \vdash$$
$$\neg(p \to p) \vee p, \neg p \vdash$$
$$\neg(p \to p) \vee p, \neg(\neg p \vee p) \vee \neg p \vdash$$
$$\neg(p \to p) \vee p, \neg(p \to p) \vee \neg p \vdash$$

Example 5. A refutation[2] of $p \to q$:

$$\neg p, p \wedge \neg q \vdash \qquad q, p \wedge \neg q \vdash$$
$$\neg p \vee q, p \wedge \neg q \vdash$$
$$p \to q, p \wedge \neg q \vdash$$
$$p \to q, \neg(p \to q) \vdash$$

The following holds:

Corollary 4.

1. *If C has a $\mathsf{HI}^{\mathsf{CPL}}$-proof, then C is valid.*
2. *If C has a $\mathsf{HI}^{\mathsf{CPL}}$-refutation[1], then C is inconsistent.*
3. *If C has a $\mathsf{HI}^{\mathsf{CPL}}$-refutation[2], then C is contingent.*

Proof. Immediately from Theorem 2 and definitions 5, 6, and 7. □

3.1 The completeness issue

The system $\mathsf{HI}^{\mathsf{CPL}}$ is complete with respect to finite HI-sets.

Theorem 3 (Completeness of $\mathsf{HI}^{\mathsf{CPL}}$ w.r.t. HI-sets). *Let Y be an at least two element finite set of wffs. If Y is a HI-set, then a sequent of the form $Y \vdash$ is provable in $\mathsf{HI}^{\mathsf{CPL}}$.*

Proof. Assume that $Y \vdash$ is in the normal form. Thus all the wffs in Y are in DNF.

By the *rank* of a sequent $Y \vdash$ (in symbols: $\mathbf{r}(Y \vdash)$) we mean the number of occurrences of the disjunction connective, $\vee$, in the wffs of Y.

Assume that Y is a finite HI-set. Note that Y has at least two elements.

Suppose that $\mathbf{r}(Y \vdash) = 0$. In this case, $Y \vdash$ is an axiom of the system.

Suppose that $\mathbf{r}(Y \vdash) > 0$. Let $\mathbf{r}(Y \vdash) = n$.

Inductive hypothesis. If $\mathbf{r}(X \vdash) < n$ and X is a HI-set of wffs in DNF, then the sequent $X \vdash$ is provable in $\mathsf{HI}^{\mathsf{CPL}}$.

If $\mathbf{r}(Y \vdash) = n$, where $n > 0$, the sequent $Y \vdash$ can be displayed as:

$$A_1, \ldots, A_{j-1}, B_1 \vee \ldots \vee B_k, A_{j+1}, \ldots, A_m \vdash$$

where $B_1, \ldots, B_k$ are elementary conjunctions and $k > 1$. As Y is a HI-set, at least one of $B_1, \ldots, B_k$ is consistent.

Let B_i be a consistent element of $\{B_1, \ldots, B_k\}$. Consider the sets Y' and Y'' defined by:

$$Y' = \{A_1, \ldots, A_{j-1}, B_i, A_{j+1}, \ldots, A_m\}$$
$$Y'' = \{A_1, \ldots, A_{j-1}, B_1 \vee \ldots \vee B_{i-1} \vee B_{i+1} \vee \ldots \vee B_k, A_{j+1}, \ldots, A_m\}$$

Clearly, $\mathbf{r}(Y' \vdash) < n$ and $\mathbf{r}(Y'' \vdash) < n$. Both $Y' \vdash$ and $Y'' \vdash$ are in the normal form. If Y is a HI-set, so is Y'. Thus, by the inductive hypothesis, the sequent $Y' \vdash$ is provable.

As for the sequent $Y'' \vdash$, there are two cases to be considered.

Case 1. $B_1 \vee \ldots \vee B_{i-1} \vee B_{i+1} \vee \ldots \vee B_k$ is consistent. Thus Y'' is a HI-set. Hence, by the inductive hypothesis, the sequent $Y'' \vdash$ is provable. But one can get $Y \vdash$ from $Y' \vdash$ and $Y'' \vdash$ by an application of rule R_1 and then, if necessary, of rule R_2.

Case 2. $B_1 \vee \ldots \vee B_{i-1} \vee B_{i+1} \vee \ldots \vee B_k$ is inconsistent. Thus all the disjuncts (of the just considered disjunction) are inconsistent. If follows that B_i is CPL-equivalent to A_j. (Clearly we have $B_i \models A_j$. But, as all the disjuncts of $B_1 \vee \ldots \vee B_{i-1} \vee B_{i+1} \vee \ldots \vee B_k$ are inconsistent, their negations are valid and hence from $A_j \models B_1 \vee \ldots \vee B_k$ we get $A_j \models B_i$.) Thus one can get $Y \vdash$ from $Y' \vdash$ by R_2.

Now assume that $Y \vdash$ is not in the normal form. In order to complete the proof it suffices to observe that each CPL-wff is CPL-equivalent to a wff in DNF and thus one can always reach a wff from its DNF-counterpart by applying rule R_2. $\square$

As a consequence of Theorem 3, Corollary 4, and definitions 5, 6, 7 one gets:

Theorem 4.

1. *A wff C is valid iff C has a $\mathsf{HI}^{\mathsf{CPL}}$-proof.*
2. *A wff C is inconsistent iff C has a $\mathsf{HI}^{\mathsf{CPL}}$-refutation*[1].
3. *A wff C is contingent iff C has a $\mathsf{HI}^{\mathsf{CPL}}$-refutation*[2].

3.2 Final remarks

The methodology used in the construction of the system $\mathsf{HI}^{\mathsf{CPL}}$, and the basic idea of the completeness proof, are very much alike to the methodology and idea applied, for different purposes, in Skura and Wiśniewski (2015).

As for this paper, the homogeneity effect has been achieved by using the notion of HI-set as a conceptual unifier. It is worth to note that the concept of minimally inconsistent set could have been used for this purpose as well. A set of wffs X is a minimally inconsistent set (MI-set for short) iff X is inconsistent, but each proper subset of X is consistent. When one deals with CPL, inconsistency, validity and contingency of wffs are expressible in terms of MI-sets as follows:

- A wff C is inconsistent iff $\{C\}$ is a MI-set.
- A wff C is valid iff $\{\neg C\}$ is a MI-set.
- A wff C is contingent iff $\{C, \neg C\}$ is a MI-set.

Thus once we have a system which "calculates" MI-sets, we get an alternative solution. A system of this kind exists (cf. Wiśniewski (2019)). The pros and cons issue remains to be studied.

The last remark is this. As for classical logic and some non-classical logics, one can define entailment by the clause:

(#) X entails A iff the set $X \cup \{\neg A\}$ is inconsistent.

However, a set of wffs can be inconsistent in different ways. One can differentiate between holistic inconsistency, minimal inconsistency, plain inconsistency, and so forth. Given this, one can then define different kinds of entailment, depending on the kind of inconsistency involved. In particular, if 'inconsistent' were replaced in (#) above with 'holistically inconsistent', we would get a non-Tarskian consequence relation with interesting properties. The system $\mathsf{HI}^{\mathsf{CPL}}$ offers a proof-theoretic account of entailment defined in this way (for the classical propositional case). However, this is another story.

References

Skura T (2009) A refutation theory. Logica Universalis 3(2):293–302
Skura T (2011) Refutation systems in propositional logic. In: Handbook of Philosophical Logic, Volume 16, Springer, pp 115–157
Skura T, Wiśniewski A (2015) A system for proper multiple-conclusion entailment. Logic and Logical Philosophy 24(2):241–253
Wiśniewski A (2019) Transmission of truth, entailment, and minimality. (submitted)

On Sequent-Type Rejection Calculi for Many-Valued Logics*

Mihail Bogojeski and Hans Tompits

Abstract Both *many-valued logics* and the notion of *axiomatic rejection* have famously been originated in modern logic by Jan Łukasiewicz during the early part of the twentieth century. The idea of an axiomatic treatment of rejection is to provide means to deduce invalid sentences from already established invalid ones. Several rejection calculi for a variety of logics have been developed, usually in either Hilbert-style or sequent-style format. In particular, previous work on rejection systems for many-valued logics include investigations by Bryll and Maduch on complete and uniform Hilbert-style rejection methods, results for infinite-valued logics by Skura, and a sequent-style rejection calculus for three-valued logics by Oetsch and Tompits. In this paper, we introduce a uniform method for constructing sequent-style rejection calculi for any given propositional finitely many-valued logic defined by means of a truth-functional semantics. The sequents we use are *many-sided*, following the well-known approach of Rousseau, which generalise standard two-sided sequents as introduced by Gentzen. Moreover, the overall construction for obtaining our rejection calculi adopts methods as discussed by Zach for uniformly constructing assertional many-sided sequent systems.

Key words: sequent-type systems, many-valued logics, proof theory, axiomatic rejection

---

Mihail Bogojeski
Institute of Software Engineering and Theoretical Computer Science
Technische Universität Berlin
e-mail: mihail.bogojeski@campus.tu-berlin.de

Hans Tompits
Institute for Logic and Computation
Technische Universität Wien
e-mail: tompits@kr.tuwien.ac.at

* The authors would like to thank Urszula Wybraniec-Skardowska and Sopo Pkhakadze for their highly appreciated help and valuable comments during the preparation of this paper.

1 Introduction

The very roots of *many-valued logics* as well as of the notion of *rejection* can be ascribed to the forefather of European logic in general, Aristotle. In *De Interpretatione* ("On Interpretation"), he suggests to assign future contingent propositions like "there will be a sea battle tomorrow", which cannot be evaluated by truth and falsity alone, a third logical status, and in the discussion of his syllogistic, one method for demonstrating the invalidity of certain syllogisms he employs is to reject arguments by reducing them to other already rejected ones. In modern logic, the subject of many-valued logics was first addressed by Jan Łukasiewicz in his 1920 paper *O logice trójwartościowej* ("On Three-valued logic") (Łukasiewicz, 1920)—in the same year that Emil Post dealt with this topic too (Post, 1920)—and he introduced in turn the term "rejection" in his 1921 paper *Logika dwuwartościowa* ("Two-valued logic") (Łukasiewicz, 1921) in which he states that by doing so he follows Brentano. An axiomatic treatment of rejection was afterwards discussed in his study of expressing Aristotle's syllogistic in modern logic (Łukasiewicz, 1939, 1957) where he introduced a Hilbert-type rejection system.[2] This was continued by his student Jerzy Słupecki (1948) and subsequently extended to a theory of rejected propositions (Słupecki, 1959; Wybraniec-Skardowska, 1969; Bryll, 1969; Słupecki et al, 1971, 1972).[3] Furthermore, axiomatic rejection methods where not only studied for classical logic (Tiomkin, 1988; Bonatti, 1993; Goranko, 1994) but also for varieties of logics, like intuitionistic logic (Skura, 1989; Dutkiewicz, 1989; Pinto and Dyckhoff, 1995; Skura, 1999, 2011), modal logics (Goranko, 1994; Skura, 2013), and others (Berger and Tompits, 2014).

Concerning rejection systems for many-valued logics, while complete and uniform rejection methods based on Hilbert-style methods are quite extensively studied in the literature, like, e.g., as put forth by Bryll and Maduch (1969) and by Skura (1993), not many works exist for sequent-style rejection systems for such logics—the notable exception being the sequent system for three-valued logics by Oetsch and Tompits (2011).

In this paper, we go beyond the three-valued case and introduce a uniform method for constructing sequent-style rejection calculi for any propositional finitely many-valued logic defined by means of a truth-functional semantics.

While Oetsch and Tompits (2011) use two-sided sequents of the form $\Gamma \dashv \Delta$, referred to as *anti-sequents*, where Γ and Δ are finite sets of formulas, in the spirit of standard sequents of the form $\Gamma \vdash \Delta$ as introduced by Gerhard Gentzen (1935a,b), we use *many-sided sequents*, which are a natural generalisation for many-valued logics of the two-sided case, following the approach of Rousseau (1967). More specifically, for axiomatising the valid formulas of an m-valued logic, Rousseau uses sequents of the form $\Gamma_1 \mid \cdots \mid \Gamma_m$, where each Γ_i ($1 \leq i \leq m$), referred to as a *component* of the sequent, is a finite set of formulas with the intuitive meaning that each component represents a truth value. That is to say, $\Gamma_1 \mid \cdots \mid \Gamma_m$ is true under an interpretation I

---

[2] Wybraniec-Skardowska (2005) refers to Łukasiewicz's system as being *biaspectual*.
[3] For an excellent survey on the development of this notion, cf., e.g., the paper by Wybraniec-Skardowska (2005).

iff there is at least one i such that, for some formula $A \in \Gamma_i$, A is assigned the i-th truth value under I. Hence, a classical sequent $\Gamma \vdash \Delta$ corresponds to $\Gamma \mid \Delta$, where the first component represents the truth value 0 while the second the truth value 1. Accordingly, we introduce *many-sided anti-sequents*, which are tuples of the form $\Gamma_1 \parallel \cdots \parallel \Gamma_m$ with a complementary semantics, i.e., an interpretation I satisfies an anti-sequent $\Gamma_1 \parallel \cdots \parallel \Gamma_m$ (i.e., *refutes* the anti-sequent) exactly in case the corresponding sequent $\Gamma_1 \mid \cdots \mid \Gamma_m$ is not true under I.

The construction of our calculi adopts a method by Zach (1993) who provides a systematic construction of many-sided sequent calculi for axiomatising validity in a given many-valued logic. This construction is based on *partial normal forms* which is a well-known method for reducing many-valued logics to two valued logic, following Rosser and Turquette (1952). Intuitively, a partial normal form defines the conditions under which a truth function takes a certain truth value. For our calculi, we introduce the notion of a *complementary partial normal form* which specifies when a truth function *does not* take a certain truth value.

2 Preliminaries

For an m-valued logic $\mathcal{L}_m$, we define

$$\mathcal{V}_m = \left\{ \frac{k}{m-1} \mid 0 \leq k \leq m-1 \right\}$$

as the set of m *truth values* of $\mathcal{L}_m$. The truth values are assumed to be ordered according to their order when interpreted as rational numbers. We use v_i, where $1 \leq i \leq m$, to denote the *i-th truth value* of $\mathcal{L}_m$, i.e., $v_i = \frac{i-1}{m-1}$.

The syntax of $\mathcal{L}_m$ is defined over an alphabet $\mathcal{A}$ consisting of (i) a countably infinite set of *propositional constants*, (ii) a collection of n-ary ($n \geq 1$) logical connectives, and (iii) *truth constants* $V_1, \ldots, V_m$, corresponding to the truth values $v_1, \ldots, v_m$, respectively.

Formulas of $\mathcal{L}_m$ are inductively defined as follows:

(P_1) Every propositional constant and every truth constant of $\mathcal{A}$ is a formula.
(P_2) If $A_1, \ldots, A_n$ are formulas and $\circ$ is an n-ary connective of $\mathcal{A}$, then $\circ(A_1, \ldots, A_n)$ is a formula.
(P_3) Formulas are constructed only according to (P_1) and (P_2).

To increase readability, binary connectives are written infix.

For an m-valued logic $\mathcal{L}_m$ over $\mathcal{A}$, an *interpretation* I is a mapping assigning each propositional constant from $\mathcal{A}$ a truth value from $\mathcal{V}_m$. Furthermore, for every n-ary connective $\circ$, we assume a corresponding *truth function*, $def_\circ^{\mathcal{L}_m}$, which is a mapping from n truth values of $\mathcal{V}_m$ to a single truth value of $\mathcal{V}_m$. Since we deal with finite-valued logics, any truth function for an n-ary connective $\circ$ can be described by a *truth table* with m^n entries.

~		∨	f	i	t		∧	f	i	t		→$_G$	f	i	t
f	t	f	f	i	t		f	f	f	f		f	t	t	t
i	f	i	i	i	t		i	f	i	i		i	f	t	t
t	f	t	t	t	t		t	f	i	t		t	f	i	t

Fig. 1: Truth tables for the connectives of $\mathcal{G}_3$.

Given an interpretation I of an m-valued logic $\mathcal{L}_m$, by a *valuation under I* we understand a mapping $val_I^{\mathcal{L}_m}(\cdot)$ which assigns to each formula A of $\mathcal{L}_m$ a truth value of $\mathcal{V}_m$ subject to the following conditions:

(i) if $A = \mathsf{V}_i$, for some truth constant V_i ($1 \leq i \leq m$), then $val_I^{\mathcal{L}_m}(A) = \mathsf{v}_i$;
(ii) if A is a propositional constant of $\mathcal{L}_m$, then $val_I^{\mathcal{L}_m}(A) = I(A)$; and
(iii) if $A = \circ(A_1, \ldots, A_n)$, for formulas $A_1, \ldots, A_n$, then

$$val_I^{\mathcal{L}_m}(A) = def_\circ^{\mathcal{L}_m}(val_I^{\mathcal{L}_m}(A_1), \ldots, val_I^{\mathcal{L}_m}(A_n)).$$

For defining a *model* of a formula, we assume a set $\mathcal{V}_m^+ \subset \mathcal{V}_m$ of *designated truth values*, which usually contains at least the truth value 1 whilst it does not contain the truth value 0. Then, an interpretation I is a model of a formula A in $\mathcal{L}_m$ if $val_I^{\mathcal{L}_m}(A) \in \mathcal{V}_m^+$. If I is not a model of A, then I is called a *countermodel* of A. A formula A is *satisfiable* if it has at least one model, and A is *valid* if every interpretation is a model of A. A valid formula is also called a *tautology*.

An interpretation I is a model of a set Γ of formulas if I is a model of every formula $A \in \Gamma$, otherwise it is a countermodel of Γ. A set Γ of formulas is satisfiable if it has at least one model.

Finally, for a logic $\mathcal{L}_m$, a formula A is a *logical consequence* of a set of formulas Γ (denoted by $\Gamma \models_{\mathcal{L}_m} A$), or Γ *entails* A, if every model of Γ is also a model of A. Given two sets Γ and Δ of formulas, we say that Γ entails Δ (symbolically, $\Gamma \models_{\mathcal{L}_m} \Delta$) if every model of Γ is a model of some $A \in \Delta$.

For illustrating our subsequent general construction of rejection systems for many-valued logics on a specific logic, we will make use of Gödel's three-valued logic (Gödel, 1932), $\mathcal{G}_3$, which is defined as follows: The syntax of $\mathcal{G}_3$ uses the four basic operators $\wedge$ (*conjunction*), $\vee$ (*disjunction*), $\to_G$ (*implication*), and $\sim$ (*negation*). For the sake of a better readability, we write the three truth values $0, \frac{1}{2}$, and 1 as **f**, **i**, and **t**, respectively. Moreover, the set of designated truth values for $\mathcal{G}_3$ consists only of the truth value **t**, i.e., $\mathcal{V}_3^+ = \{\mathbf{t}\}$. The semantics of the four $\mathcal{G}_3$ connectives $\sim, \vee, \wedge$, and $\to_G$ is defined by means of the truth tables depicted in Figure 1.

3 Many-Sided Anti-Sequents

Definition 3.1. *For an m-valued logic $\mathcal{L}_m$, a* many-sided anti-sequent, *or simply, an* anti-sequent, *is an ordered tuple of the form $\mathfrak{R} = \Gamma_1 \parallel \cdots \parallel \Gamma_m$, where each Γ_i is a finite set of formulas of $\mathcal{L}_m$, called the i-th component of $\mathfrak{R}$ $(1 \leq i \leq m)$.*

Intuitively, the i-th component of an anti-sequent corresponds to the i-th truth value $v_i \in \mathcal{V}_m$ $(1 \leq i \leq m)$. The components in $\mathfrak{R}$ are also ordered ascendingly, where the ordering of the components is determined by the ordering of the corresponding truth values. Given a formula A and a component Γ, following custom, we write "Γ, A" instead of $\Gamma \cup \{A\}$. The *empty m-component many-sided anti-sequent* is given by the m-tuple $\mathfrak{T}_m = \emptyset \parallel \cdots \parallel \emptyset$, i.e., in which all components coincide with the empty set.

For two m-component anti-sequents $\mathfrak{R}_1 = \Gamma_1 \parallel \cdots \parallel \Gamma_m$ and $\mathfrak{R}_2 = \Delta_1 \parallel \cdots \parallel \Delta_m$, we define the combination of $\mathfrak{R}_1$ and $\mathfrak{R}_2$ by

$$\mathfrak{R}_1, \mathfrak{R}_2 = \Gamma_1, \Delta_1 \parallel \cdots \parallel \Gamma_m, \Delta_m.$$

For an anti-sequent $\mathfrak{R} = \Gamma_1 \parallel \cdots \parallel \Gamma_m$ and a set Δ of formulas,

$$\mathfrak{R}, [i : \Delta]$$

denotes the many-sided sequent that has the same components as $\mathfrak{R}$ but additionally contains Δ in its i-th component, i.e.,

$$\mathfrak{R}, [i : \Delta] = \Gamma_1 \parallel \cdots \parallel \Gamma_i, \Delta \parallel \cdots \parallel \Gamma_m.$$

This notation can also be applied repeatedly to an anti-sequent in the following manner: Let $\mathfrak{R} = \Gamma_1 \parallel \cdots \parallel \Gamma_m$, then

$$\mathfrak{R}, [i_1 : \Delta_1], \ldots, [i_n : \Delta_n] = \Gamma_1 \parallel \cdots \parallel \Gamma_{i_1}, \Delta_1 \parallel \cdots \parallel \Gamma_{i_n}, \Delta_n \parallel \cdots \parallel \Gamma_m.$$

Given an anti-sequent $\mathfrak{R}$, a set of formulas Δ, and a set $M \subseteq \{1, \ldots, m\}$, we define

$$\mathfrak{R}, [M : \Delta] = \mathfrak{R}, [i_1 : \Delta], \ldots, [i_n : \Delta],$$

where $M = \{i_1, \ldots, i_n\}$. For example, consider the three-component anti-sequent $\mathfrak{R} = \Gamma_1 \parallel \Gamma_2 \parallel \Gamma_3$, an arbitrary set Δ of formulas, and $M = \{1, 3\}$. Then, $\mathfrak{R}, [M : \Delta] = \mathfrak{R}, [1 : \Delta], [3 : \Delta] = \Gamma_1, \Delta \parallel \Gamma_2 \parallel \Gamma_3, \Delta$.

The semantics of anti-sequents is defined as follows:

Definition 3.2. *Let I be an interpretation of an m-valued logic $\mathcal{L}_m$ and $\mathfrak{R} = \Gamma_1 \parallel \cdots \parallel \Gamma_m$ an m-component anti-sequent. Then, I* refutes *$\mathfrak{R}$ if, for every i $(1 \leq i \leq m)$ and every formula $A \in \Gamma_i$, $\mathrm{val}_I^{\mathcal{L}_m}(A) \neq \mathsf{v}_i$. In this case, I is said to be a* countermodel *of $\mathfrak{R}$.*

An anti-sequent $\mathfrak{R}$ is *refutable* if there is at least one interpretation I that refutes $\mathfrak{R}$. Furthermore, $\mathfrak{R}$ is *unsatisfiable* if every interpretation refutes $\mathfrak{R}$. Note that the empty anti-sequent $\mathfrak{T}_m$ is refuted by every interpretation.

In contrast to two-sided sequents, many-sided sequents do not directly encode entailment but it can be expressed in the following way:

Theorem 3.3. *Let $\mathcal{L}_m$ be an m-valued logic, and Γ and Δ finite sets of formulas in $\mathcal{L}_m$. Additionally, let $M^+ = \{i \mid v_i \in \mathcal{V}_m^+\}$ and $M^- = \{i \mid v_i \in \mathcal{V}_m \setminus \mathcal{V}_m^+\}$. Then, it holds that $\Gamma \not\models_{\mathcal{L}_m} \Delta$ iff the many-sided anti-sequent $\mathfrak{T}_m$, $[M^- : \Gamma]$, $[M^+ : \Delta]$ is refutable.*

4 Encoding Truth Functions in Two-Valued Logic

As a preparatory step towards our rejection calculi for many-valued logics, we now discuss methods for reducing many-valued logics to two valued logic based on the concept of a *partial normal form*. More specifically, following Rosser and Turquette (1952), the *i-th partial normal form* of a formula $\circ(A_1, \ldots, A_n)$ is a conjunctive-normal form specification in two-valued logic of the conditions under which the truth function $def_\circ^{\mathcal{L}_m}$ takes the truth value v_i. Given this, we define the *i-th complementary partial normal form* of a formula $\circ(A_1, \ldots, A_n)$ to describe the conditions under which $def_\circ^{\mathcal{L}_m}$ does *not* take the truth value v_i. The complementary partial normal forms are then used to define the inference rules of the rejection calculus for $\mathcal{L}_m$.

The construction of the partial normal forms is in turn based on the notion of a *signed formula expression* as discussed by Zach (1993).

4.1 Signed Formula Expressions

By a *signed formula* we understand an expression of the form A^w, where w is a truth value and A is a formula. A *signed formula expression* (SFE) is a formula built up from signed formulas using the connectives $\dot{\vee}$, $\dot{\wedge}$, and $\dot{\neg}$. A signed formula expression of the form A^w or $\dot{\neg}A^w$ is called a *signed literal*. It is called an *atomic literal* iff A is atomic. In that case, A^w is also called a *signed atom*.

The relation $\models_{SFE}$ between an interpretation I and a signed formula expression F, composed of formulas of an m-valued logic $\mathcal{L}_m$, is inductively defined as follows:

(i) if $F = A^w$, where A is a formula and $w \in \mathcal{V}_m$, then $I \models_{SFE} F$ iff $val_I^{\mathcal{L}_m}(A) = w$;
(ii) if $F = \dot{\neg}G$, for an SFE G, then $I \models_{SFE} F$ iff $I \not\models_{SFE} G$;
(iii) if $F = G \dot{\wedge} H$, for SFEs G and H, then $I \models_{SFE} F$ iff $I \models_{SFE} G$ and $I \models_{SFE} H$; and
(vi) if $F = G \dot{\vee} H$, for SFEs G and H, then $I \models_{SFE} F$ iff $I \models_{SFE} G$ or $I \models_{SFE} H$.

If $I \models_{SFE} F$ holds, we say that *I satisfies F*. A signed formula expression F is *satisfiable* iff there is an interpretation I such that $I \models_{SFE} F$. Furthermore, F is *valid* iff every interpretation satisfies F. Two signed formula expressions F and G are *equivalent*, symbolically $F \equiv G$, iff, for every interpretation I, $I \models_{SFE} F$ precisely in case $I \models_{SFE} G$ holds.

Assume that the symbol $\dot\neg$ always occurs directly before the signed formulas of an SFE F. Then, F is called *positive* if all of its signed literals have the form A^w and F is *negative* if every signed literal in F has the form $\dot\neg A^w$.

We employ the following abbreviations for SFEs: we use $\dot\bigvee_{i=1}^{n} F_i$ to stand for $F_1 \dot\vee (F_2 \dot\vee \cdots (F_{n-1} \dot\vee F_n) \cdots)$ and $\dot\bigwedge_{i=1}^{n} F_i$ for $F_1 \dot\wedge (F_2 \dot\wedge \cdots (F_{n-1} \dot\wedge F_n) \cdots)$. If $n = 0$, we obtain the *empty disjunction* (which always evaluates to 0) and the *empty conjunction* (which always evaluates to 1), respectively.

Parentheses can be omitted in the standard way: binary operators associate to the right, and precedence is given in the decreasing order $\dot\neg, \dot\wedge$, and $\dot\vee$. Since the semantics of the operators $\dot\neg, \dot\wedge$, and $\dot\vee$ is similar to the semantics of the classical two-valued operators $\neg, \wedge$, and $\vee$, standard equivalences in classical logic such as the associative, commutative, distributive, and de Morgan's laws also hold for SFEs.

Proposition 4.1 (Zach, 1993). *Let A be a formula of an m-valued logic $\mathcal{L}_m$. Then, the following statements are equivalent: (i) A is a tautology; (ii) the SFE $\dot\bigvee_{w \in V_m^+} A^w$ is valid; (iii) the SFE $\dot\bigwedge_{w \in V_m \setminus V_m^+} \dot\neg A^w$ is valid; (iv) the SFE $\dot\bigvee_{w \in V_m \setminus V_m^+} A^w$ is unsatisfiable; and (v) the SFE $\dot\bigwedge_{w \in V_m^+} \dot\neg A^w$ is unsatisfiable.*

Proposition 4.2 (Zach, 1993). *Every SFE F can be transformed into an equivalent positive SFE $p(F)$ as well as into an equivalent negative SFE $n(F)$.*

Finally, anti-sequents can also be expressed in terms of SFEs as follows:

Theorem 4.1. *Let I be an interpretation and $\mathfrak{R} = \Gamma_1 \parallel \cdots \parallel \Gamma_m$ an m-component many-sided anti-sequent, where $\Gamma_i = \{A_{i,1}, \ldots, A_{i,k_i}\}$, for $1 \le i \le m$. Then, I refutes $\mathfrak{R}$ iff it satisfies the SFE*

$$\dot\bigwedge_{j_1=1}^{k_1} \dot\neg A_{1,j_1}^{v_1} \dot\wedge \cdots \dot\wedge \dot\bigwedge_{j_m=1}^{k_m} \dot\neg A_{m,j_m}^{v_m}.$$

4.2 Partial Normal Forms

Let $\circ(A_1, \ldots, A_n)$ be a formula of an m-valued logic $\mathcal{L}_m$, where $\circ$ is an n-ary connective. Following Rosser and Turquette (1952), the i-th partial normal form (PNF) for $\circ(A_1, \ldots, A_n)$, denoted by $pnf_i[\circ(A_1, \ldots, A_n)]$, is constructed from the truth table of $\circ$ in the following way:

(i) We find the tuples of truth values for which the truth function $def_\circ^{\mathcal{L}_m}$ does not take the truth value v_i:

$$W_i = \{(w_1, \ldots, w_n) \mid def_\circ^{\mathcal{L}_m}(w_1, \ldots, w_n) \ne v_i\}.$$

(ii) We construct $pnf_i[\circ(A_1, \ldots, A_n)]$ by taking the complement of the situations described in W_i and linking them conjunctively:

$$pnf_i[\circ(A_1,\ldots,A_n)] =$$
$$\dot{\bigwedge}_{(w_1,\ldots,w_n)\in W_i} \left(\dot{\bigvee}_{u_1\in \mathcal{V}_m\setminus\{w_1\}} A_1^{u_1} \dot{\vee} \cdots \dot{\vee} \dot{\bigvee}_{u_n\in \mathcal{V}_m\setminus\{w_n\}} A_n^{u_n} \right).$$

Then, the following holds:

Proposition 4.3 (Rosser and Turquette, 1952; Zach, 1993). *For a given formula $\circ(A_1,\ldots,A_n)$ in an m-valued logic $\mathcal{L}_m$, where $\circ$ is an n-ary connective, the i-th partial normal form $pnf_i[\circ(A_1,\ldots,A_n)]$ is equivalent to the SFE $(\circ(A_1,\ldots,A_n))^{v_i}$.*

Now we can obtain the corresponding complementary partial normal form of a given PNF by negating the PNF and using de Morgan's laws to push the negations directly before each signed formula.

Definition 4.2. *Given the i-th partial normal form $pnf_i[\circ(A_1,\ldots,A_n)]$ of a formula $\circ(A_1,\ldots,A_n)$, where $\circ$ is an n-ary connective, the i-th complementary partial normal form (CPNF) for $\circ(A_1,\ldots,A_n)$, denoted by $cpnf_i[\circ(A_1,\ldots,A_n)]$, is given as follows:*

$$cpnf_i[\circ(A_1,\ldots,A_n)] =$$
$$\dot{\bigvee}_{(w_1,\ldots,w_n)\in W_i} \left(\dot{\bigwedge}_{u_1\in \mathcal{V}_m\setminus\{w_1\}} \dot{\neg} A_1^{u_1} \dot{\wedge} \cdots \dot{\wedge} \dot{\bigwedge}_{u_n\in \mathcal{V}_m\setminus\{w_n\}} \dot{\neg} A_n^{u_n} \right).$$

Theorem 4.3. *For a given formula $\circ(A_1,\ldots,A_n)$ in an m-valued logic $\mathcal{L}_m$, where $\circ$ is an n-ary connective, the i-th complementary partial normal form $cpnf_i[\circ(A_1,\ldots,A_n)]$ is equivalent to the SFE $\dot{\neg}(\circ(A_1,\ldots,A_n))^{v_i}$.*

Proof. This is a direct consequence of Proposition 4.3 and the fact that

$$cpnf_i[\circ(A_1,\ldots,A_n)] \equiv \dot{\neg} pnf_i[\circ(A_1,\ldots,A_n)].$$

$\square$

Clearly, the method for constructing PNFs as described above does in general not yield the shortest SFEs possessing the properties of $pnf_i[\cdot]$ and $cpnf_i[\cdot]$ as described by Proposition 4.3 and Theorem 4.3, respectively. However, standard methods for minimising Boolean formulas, such as the Quine-McCluskey procedure, can be used to obtain minimal PNFs. The use of "minimal" here is meant with respect to the number of conjuncts and the number of formulas per conjuncts.

Example 4.4. *Figure 2 depicts minimal CPNFs for Gödel's three valued logic $\mathcal{G}_3$ obtained from the truth tables in Figure 1. We will use these for providing the explicit rules of the complementary calculus for $\mathcal{G}_3$. An interesting case to observe here is the second CPNF for the negation connective $\sim$: Since $def_{\sim}^{\mathcal{G}_3}$ never takes the truth value* i, *the corresponding PNF is unsatisfiable and thus the CPNF is always true, which makes the normal forms very simple. However, the general construction of the PNF would yield 3^2 conjuncts, which is as large as a PNF for a binary connective in a three-valued logic can possibly get.*

$\dot\neg(\sim A)^{\mathbf{f}} = \dot\neg A^{\mathbf{t}} \dot\wedge \dot\neg A^{\mathbf{i}};$

$\dot\neg(\sim A)^{\mathbf{i}} = \top;$

$\dot\neg(\sim A)^{\mathbf{t}} = \dot\neg A^{\mathbf{f}};$

$\dot\neg(A \vee B)^{\mathbf{f}} = \dot\neg A^{\mathbf{f}} \dot\vee \dot\neg B^{\mathbf{f}};$

$\dot\neg(A \vee B)^{\mathbf{i}} = (\dot\neg A^{\mathbf{i}} \dot\wedge \dot\neg B^{\mathbf{i}}) \dot\vee (\dot\neg A^{\mathbf{i}} \dot\wedge \dot\neg A^{\mathbf{f}}) \dot\vee (\dot\neg B^{\mathbf{i}} \dot\wedge \dot\neg B^{\mathbf{f}});$

$\dot\neg(A \vee B)^{\mathbf{t}} = \dot\neg A^{\mathbf{t}} \dot\wedge \dot\neg B^{\mathbf{t}};$

$\dot\neg(A \wedge B)^{\mathbf{f}} = \dot\neg A^{\mathbf{f}} \dot\wedge \dot\neg B^{\mathbf{f}};$

$\dot\neg(A \wedge B)^{\mathbf{i}} = (\dot\neg A^{\mathbf{i}} \dot\wedge \dot\neg B^{\mathbf{i}}) \dot\vee (\dot\neg A^{\mathbf{i}} \dot\wedge \dot\neg A^{\mathbf{t}}) \dot\vee (\dot\neg B^{\mathbf{i}} \dot\wedge \dot\neg B^{\mathbf{t}});$

$\dot\neg(A \wedge B)^{\mathbf{t}} = \dot\neg A^{\mathbf{t}} \dot\vee \dot\neg B^{\mathbf{t}};$

$\dot\neg(A \to_G B)^{\mathbf{f}} = (\dot\neg A^{\mathbf{t}} \dot\wedge \dot\neg A^{\mathbf{i}}) \dot\vee \dot\neg B^{\mathbf{f}};$

$\dot\neg(A \to_G B)^{\mathbf{i}} = \dot\neg A^{\mathbf{t}} \dot\vee \dot\neg B^{\mathbf{i}};$

$\dot\neg(A \to_G B)^{\mathbf{t}} = (\dot\neg A^{\mathbf{f}} \dot\wedge \dot\neg A^{\mathbf{i}} \dot\wedge \dot\neg B^{\mathbf{t}}) \dot\vee (\dot\neg A^{\mathbf{f}} \dot\wedge \dot\neg B^{\mathbf{i}} \dot\wedge \dot\neg B^{\mathbf{t}}).$

Fig. 2: The minimal CPNFs for the $\mathcal{G}_3$ connectives $\sim$, $\vee$, $\wedge$, and $\to_G$.

5 Many-Sided Anti-Sequent Calculi

5.1 Postulates of the Calculi

Recall that the i-th complementary partial normal form for a formula $\circ(A_1, \ldots, A_n)$ describes the conditions under which the formula does not take the truth value v_i. We exploit this property for obtaining the inference rules for each connective $\circ$.

To begin with, we give a result which shows how a negative conjunctive SFE can be converted into an equivalent many-sided anti-sequent.

Theorem 5.1. *Let F be a negative conjunctive SFE of the form* $\dot\neg A_1^{w_1} \dot\wedge \cdots \dot\wedge \dot\neg A_n^{w_n}$, *where* $A_1, \ldots, A_n$ *are formulas of the m-valued logic* $\mathcal{L}_m$ *and* $w_1, \ldots, w_n \in \mathcal{V}_m$. *Furthermore, for each* $i \in \{1, \ldots, m\}$, *let* $\Gamma_i = \{A \mid \dot\neg A^w \in literals(F), w = v_i\}$, *where* $literals(F)$ *denotes the set of all signed literals that are used in F.*
Then, for any m-valued interpretation I, I $\models_{SFE} F$ *iff I refutes* $\mathfrak{R} = \Gamma_1 \parallel \cdots \parallel \Gamma_m$.

Proof. The result follows by straightforward arguments in view of the construction of each Γ_i ($1 \leq i \leq m$) as well as from the definition of $\models_{SFE}$ and the semantics of anti-sequents. □

Given this result, we can construct the inference rules as follows:

Definition 5.2. *Let* $cpnf_i[\circ(A_1, \ldots, A_n)] = \dot\bigvee_{j=1}^{k_i} \sigma^j_{\circ:v_i}(A_1, \ldots, A_n)$ *be the i-th CPNF of a formula* $\circ(A_1, \ldots, A_n)$, *where* $\circ$ *is an n-ary connective,* k_i *is the number of disjuncts in the CPNF, and* $\sigma^j_{\circ:v_i}(A_1, \ldots, A_n)$ *is the j-th disjunct of the CPNF, which is a conjunction of negated signed atoms. Furthermore, let* $\mathfrak{R}$ *be an arbitrary m-component anti-sequent*

$$\frac{\Gamma \parallel \Delta, A \parallel \Pi, A}{\Gamma, \sim A \parallel \Delta \parallel \Pi} \; (\sim: \mathbf{f}) \qquad \frac{\Gamma \parallel \Delta \parallel \Pi}{\Gamma \parallel \Delta, \sim A \parallel \Pi} \; (\sim: \mathbf{i}) \qquad \frac{\Gamma, A \parallel \Delta \parallel \Pi}{\Gamma \parallel \Delta \parallel \Pi, \sim A} \; (\sim: \mathbf{t})$$

$$\frac{\Gamma, A \parallel \Delta \parallel \Pi}{\Gamma, A \vee B \parallel \Delta \parallel \Pi} \; (\vee: \mathbf{f}^1) \qquad \frac{\Gamma, B \parallel \Delta \parallel \Pi}{\Gamma, A \vee B \parallel \Delta \parallel \Pi} \; (\vee: \mathbf{f}^2) \qquad \frac{\Gamma \parallel \Delta \parallel \Pi, A, B}{\Gamma \parallel \Delta \parallel \Pi, A \vee B} \; (\vee: \mathbf{t})$$

$$\frac{\Gamma \parallel \Delta, A, B \parallel \Pi}{\Gamma \parallel \Delta, A \vee B \parallel \Pi} \; (\vee: \mathbf{i}^1) \qquad \frac{\Gamma, A \parallel \Delta, A \parallel \Pi}{\Gamma \parallel \Delta, A \vee B \parallel \Pi} \; (\vee: \mathbf{i}^2) \qquad \frac{\Gamma, B \parallel \Delta, B \parallel \Pi}{\Gamma \parallel \Delta, A \vee B \parallel \Pi} \; (\vee: \mathbf{i}^3)$$

$$\frac{\Gamma, A, B \parallel \Delta \parallel \Pi}{\Gamma, A \wedge B \parallel \Delta \parallel \Pi} \; (\wedge: \mathbf{f}) \qquad \frac{\Gamma \parallel \Delta \parallel \Pi, A}{\Gamma \parallel \Delta \parallel \Pi, A \wedge B} \; (\wedge: \mathbf{t}^1) \qquad \frac{\Gamma \parallel \Delta \parallel \Pi, B}{\Gamma \parallel \Delta \parallel \Pi, A \wedge B} \; (\wedge: \mathbf{t}^2)$$

$$\frac{\Gamma \parallel \Delta, A, B \parallel \Pi}{\Gamma \parallel \Delta, A \wedge B \parallel \Pi} \; (\wedge: \mathbf{i}^1) \qquad \frac{\Gamma \parallel \Delta, A \parallel \Pi, A}{\Gamma \parallel \Delta, A \wedge B \parallel \Pi} \; (\wedge: \mathbf{i}^2) \qquad \frac{\Gamma \parallel \Delta, B \parallel \Pi, B}{\Gamma \parallel \Delta, A \wedge B \parallel \Pi} \; (\wedge: \mathbf{i}^3)$$

$$\frac{\Gamma \parallel \Delta, A \parallel \Pi, A}{\Gamma, A \rightarrow_G B \parallel \Delta \parallel \Pi} \; (\rightarrow_G: \mathbf{f}^1) \qquad \frac{\Gamma, B \parallel \Delta \parallel \Pi}{\Gamma, A \rightarrow_G B \parallel \Delta \parallel \Pi} \; (\rightarrow_G: \mathbf{f}^2)$$

$$\frac{\Gamma \parallel \Delta \parallel \Pi, A}{\Gamma \parallel \Delta, A \rightarrow_G B \parallel \Pi} \; (\rightarrow_G: \mathbf{i}^1) \qquad \frac{\Gamma \parallel \Delta, B \parallel \Pi}{\Gamma \parallel \Delta, A \rightarrow_G B \parallel \Pi} \; (\rightarrow_G: \mathbf{i}^2)$$

$$\frac{\Gamma, A \parallel \Delta, A \parallel \Pi, B}{\Gamma \parallel \Delta \parallel \Pi, A \rightarrow_G B} \; (\rightarrow_G: \mathbf{t}^1) \qquad \frac{\Gamma, A \parallel \Delta, B \parallel \Pi, B}{\Gamma \parallel \Delta \parallel \Pi, A \rightarrow_G B} \; (\rightarrow_G: \mathbf{t}^2)$$

Fig. 3: Inference rules of $\mathbf{MSR}_{\mathcal{G}_3}$.

and $\hat{\sigma}^j_{\circ:v_i}(A_1, \ldots, A_n)$ the anti-sequent constructed from $\sigma^j_{\circ:v_i}(A_1, \ldots, A_n)$ as described in Theorem 5.1.

Then, for every $j \in \{1, \ldots, k_i\}$, the inference rule $(\circ : v_i^j)$ is given as follows:

$$\frac{\mathfrak{R}, \hat{\sigma}^j_{\circ:v_i}(A_1, \ldots, A_n)}{\mathfrak{R}, [i : \circ(A_1, \ldots, A_n)]} \; (\circ : v_i^j).$$

Note that in contrast to standard (assertional) sequent-type calculi (axiomatising valid sequents), the inference rules introduced above have only single premisses. Indeed, this is a general pattern in sequent-style rejection calculi: if an inference rule for standard sequents for a connective has n premisses, then there are usually n corresponding unary inference rules in the associated rejection calculus. Intuitively, what is exhaustive search in a standard sequent calculus becomes nondeterminism in a rejection calculus.

We are now in a position to formulate the postulates of our calculus:

$$\frac{P,Q \parallel \varnothing \parallel P,Q}{P,P,Q,Q \parallel \varnothing \parallel P,Q} \ (c:\mathbf{f}, c:\mathbf{f})$$
$$\frac{P,P,Q,Q \parallel \varnothing \parallel P,Q}{P,P,Q,Q \parallel \varnothing \parallel P \vee Q} \ (\vee:\mathbf{t})$$
$$\frac{P,P,Q,Q \parallel \varnothing \parallel P \vee Q}{P,Q \parallel \varnothing \parallel P \vee Q, \sim P, \sim Q} \ (\sim:\mathbf{t}, \sim:\mathbf{t})$$
$$\frac{P,Q \parallel \varnothing \parallel P \vee Q, \sim P, \sim Q}{\sim Q \to_G P, \sim P \to_G Q \parallel \varnothing \parallel P \vee Q, \sim P, \sim Q} \ (\to_G:\mathbf{f}^2, \to_G:\mathbf{f}^2)$$
$$\frac{\sim Q \to_G P, \sim P \to_G Q \parallel \varnothing \parallel P \vee Q, \sim P, \sim Q}{\sim Q \to_G P, \sim Q \to_G P \parallel \sim Q \to_G P, \sim P \to_G Q \parallel P \vee Q} \ (\to_G:\mathbf{i}^1, \to_G:\mathbf{i}^1)$$
$$\frac{\sim Q \to_G P, \sim Q \to_G P \parallel \sim Q \to_G P, \sim P \to_G Q \parallel P \vee Q}{(\sim Q \to_G P) \wedge (\sim P \to_G Q) \parallel (\sim Q \to_G P) \wedge (\sim P \to_G Q) \parallel P \vee Q} \ (\wedge:\mathbf{f}, \wedge:\mathbf{i}^1).$$

Fig. 4: A $\mathbf{MSR}_{\mathcal{G}_3}$ proof of the anti-sequent $A \parallel A \parallel B$ from Example 5.5.

Definition 5.3. *Let $\mathcal{L}_m$ be an m-valued logic and $\mathcal{V}_m = \{\mathsf{v}_1, \ldots, \mathsf{v}_m\}$ its set of truth values. Then, the many-sided rejection calculus $\mathbf{MSR}_{\mathcal{L}_m}$ consists of*

(i) *axiom schemata of the form $\mathfrak{T}_m, [M_1 : P_1], \ldots, [M_k : P_k]$, where $\mathfrak{T}_m$ is the empty anti-sequent, $P_1, \ldots, P_k$ are different propositional constants, and $M_1, \ldots, M_k \subset \{1, \ldots, m\}$; and*
(ii) *the inference rules $(\circ : \mathsf{v}_i^1), \ldots, (\circ : \mathsf{v}_i^{k_i})$ as given in Definition 5.2, for every connective $\circ$ of $\mathcal{L}_m$ and every truth value $\mathsf{v}_i \in \mathcal{V}_m$, where $1 \leq i \leq m$.*

Note that the axioms of $\mathbf{MSR}_{\mathcal{L}_m}$ are anti-sequents whose components are sets of propositional constants such that no constant appears in all components. Furthermore, since all inference rules of $\mathbf{MSR}_{\mathcal{L}_m}$ are unary, proofs are sequences of anti-sequents having a single axiom, which is in contrast to usual (assertional) sequent systems in which proofs are trees.

Example 5.4. *Consider the rejection calculus $\mathbf{MSR}_{\mathcal{G}_3}$ for Gödel's three-valued logic $\mathcal{G}_3$, which is obtained on the basis of the CPNFs given in Figure 2. The corresponding inference rules are depicted in Figure 3.*

$\mathbf{MSR}_{\mathcal{G}_3}$ can be compared to the anti-sequent calculus $\mathbf{SRCL}$ as introduced by Oetsch and Tompits (2011) which uses standard two-sided sequents. Both calculi have roughly the same number of rules, but $\mathbf{MSR}_{\mathcal{G}_3}$ can be argued to have more "elegant" postulates, since every inference rule introduces only one connective at a time whilst $\mathbf{SRCL}$ consists of standard and non-standard rules, where the latter introduce two connectives in each application (standard rules correspond to the usual sequent inference rules as for classical logic). This reflects the fact that $\mathbf{SRCL}$ is based upon a sequent calculus for three-valued logic due to Avron (1991b) which has a similar feature.

Example 5.5. *Consider the two formulas $A = (\sim Q \to_G P) \wedge (\sim P \to_G Q)$ and $B = P \vee Q$, where P and Q are distinct propositional constants. It holds that $A \not\models_{\mathcal{G}_3} B$ in view of the proof in $\mathbf{MSR}_{\mathcal{G}_3}$ of the anti-sequent $A \parallel A \parallel B$, as given in Figure 4, and by Theorem 3.3.*

5.2 Soundness and Completeness

We now show soundness and completeness of our calculi. We start with the following lemma:

Lemma 5.6. *Let I be an m-valued interpretation. If I refutes the axiom of a proof in* $\mathbf{MSR}_{\mathcal{L}_m}$, *then I refutes all anti-sequents occurring in the proof.*

Proof. It suffices to show that if I refutes the premiss of an inference rule, then it refutes the conclusion of the rule as well. So, consider an application of some rule $(\circ : v_i^j)$ as in Definition 5.2 with premiss $\mathfrak{R}, \hat{\sigma}_{\circ:v_i}^j(A_1, \ldots, A_n)$ and assume that I refutes the premiss. Then, I refutes both $\mathfrak{R}$ and $\hat{\sigma}_{\circ:v_i}^j(A_1, \ldots, A_n)$. Since I refutes $\hat{\sigma}_{\circ:v_i}^j(A_1, \ldots, A_n)$, in view of Theorem 4.1, I satisfies the SFE $\sigma_{\circ:v_i}^j(A_1, \ldots, A_n)$ and thus the whole instance of the i-th CPNF for $\circ(A_1, \ldots, A_n)$, which has the form $\bigvee_{j=1}^{k_i} \sigma_{\circ:v_i}^j(A_1, \ldots, A_n)$. Hence, by Theorem 4.3, I satisfies the SFE $\dot{\neg}(\circ(A_i, \ldots, A_n))^{v_i}$, and thus refutes the m-component anti-sequent $\mathfrak{T}_m, [i : \circ(A_1, \ldots, A_n)]$, where $\mathfrak{T}_m$ is the empty anti-sequent. As I refutes $\mathfrak{R}$, it therefore refutes $\mathfrak{R}, \mathfrak{T}_m, [i : \circ(A_1, \ldots, A_n)]$ as well. But the latter coincides with $\mathfrak{R}, [i : \circ(A_1, \ldots, A_n)]$, hence I refutes the conclusion of $(\circ : v_i^j)$. □

From this, soundness immediately follows:

Theorem 5.7. *If an anti-sequent is provable in* $\mathbf{MSR}_{\mathcal{L}_m}$, *then it is refutable in* $\mathcal{L}_m$.

Proof. Consider an anti-sequent $\mathfrak{R}$ which is provable in $\mathbf{MSR}_{\mathcal{L}_m}$ and let $\mathfrak{R}'$ be the axiom of a proof of $\mathfrak{R}$. Then, $\mathfrak{R}'$ is of the form $\mathfrak{T}_m, [M_1 : P_1], \ldots, [M_k : P_k]$, where $\mathfrak{T}_m$ is the empty anti-sequent, $P_1, \ldots, P_k$ are different propositional constants, and $M_1, \ldots, M_k \subset \{1, \ldots, m\}$. We construct an interpretation I as follows: For each propositional constant P_j, choose one $i \in \{1, \ldots, m\} \setminus M_j$ and define $val_I^{\mathcal{L}_m}(P_j) = v_i$. For this interpretation I, whenever a propositional constant is contained in an i-th component of $\mathfrak{R}'$, $val_I^{\mathcal{L}_m} \neq v_i$ holds, which means that I refutes the anti-sequent $\mathfrak{R}'$. Therefore, by Lemma 5.6, I refutes $\mathfrak{R}$ as well. □

Consequently, the unique axioms in proofs in $\mathbf{MSR}_{\mathcal{L}_m}$ allow to directly construct counterexamples for provable anti-sequents.

Now we deal with the completeness of our calculus.

Theorem 5.8. *If an anti-sequent is refutable in* $\mathcal{L}_m$, *then it is provable in* $\mathbf{MSR}_{\mathcal{L}_m}$.

Proof. The proof proceeds by the method of a *reduction tree*, as originally used by Schütte (1956). We show that every m-component many-sided anti-sequent $\mathfrak{R}$ is either provable in $\mathbf{MSR}_{\mathcal{L}_m}$ or irrefutable in $\mathcal{L}_m$ (i.e., there is no interpretation I that refutes $\mathfrak{R}$).

A reduction tree, $T_{\mathfrak{R}}$, is an upward, rooted tree of many-sided anti-sequents constructed from $\mathfrak{R}$ in stages as follows:

STAGE 0: Write $\mathfrak{R}$ at the root of the reduction tree.

STAGE $s + 1$: If the topmost many-sided anti-sequent of a branch at stage s contains only propositional constants, then this branch is called *closed* and the reduction on this branch is stopped. A branch that does not have this property is called *open*. For every open branch B at stage s, choose a non-atomic formula A contained in some i-th component of the topmost many-sided anti-sequent $\mathfrak{R}^s$ of B. Then, A has the form $\circ(A_1,\ldots,A_n)$ and $\mathfrak{R}^s = \mathfrak{R}', [i : \circ(A_1,\ldots,A_n)]$, where $\circ$ is some n-ary connective, $A_1,\ldots,A_n$ are propositional formulas, and $\mathfrak{R}'$ is some anti-sequent. Replace $\mathfrak{R}^s$ in the reduction tree by the figure

$$\frac{\mathfrak{R}', \hat{\sigma}^1_{\circ:v_i}(A_1,\ldots,A_n) \quad \cdots \quad \mathfrak{R}', \hat{\sigma}^{k_i}_{\circ:v_i}(A_1,\ldots,A_n)}{\mathfrak{R}', [i : \circ(A_1,\ldots,A_n)]},$$

where every anti-sequent $\mathfrak{R}', \hat{\sigma}^j_{\circ:v_i}(A_1,\ldots,A_n)$, $1 \leq j \leq k_i$, is the premiss of an instance of the $\mathbf{MSR}_{\mathcal{L}_m}$ inference rule $(\circ : v^j_i)$ constructed as in Definition 5.2. This concludes the construction of stage $s + 1$.

A reduction tree $T_\mathfrak{R}$ constructed in this way will always be finite, since at some stage s, every formula with a connective will be reduced to atomic formulas. Let $\mathcal{C}$ be the set of all closed branches in $T_\mathfrak{R}$. We distinguish two cases based on the branches in $\mathcal{C}$:

CASE 1: For at least one branch B in $\mathcal{C}$, its topmost many-sided anti-sequent, $\mathfrak{R}^B$, is an axiom of $\mathbf{MSR}_{\mathcal{L}_m}$, i.e., there is no propositional constant that appears in every component of $\mathfrak{R}^B$. With $\mathfrak{R}^B$ as an axiom, we can easily construct a proof for $\mathfrak{R}$ in $\mathbf{MSR}_{\mathcal{L}_m}$ by following the branch B to the root of the reduction tree $T_\mathfrak{R}$.

CASE 2: There is no closed branch B in $\mathcal{C}$ such that its topmost many-sided anti-sequent $\mathfrak{R}^B$ is an axiom in $\mathbf{MSR}_{\mathcal{L}_m}$. This means that for every branch B in $\mathcal{C}$, there is no interpretation I which refutes its topmost many-sided anti-sequent $\mathfrak{R}^B$. It can be shown by induction on the depth d of a many-sided anti-sequent in the tree $T_\mathfrak{R}$, that every anti-sequent in the reduction tree $T_\mathfrak{R}$, including the root anti-sequent $\mathfrak{R}$, is irrefutable. Here, the depth is defined as follows: Let $\mathfrak{R}^s$ be a topmost anti-sequent that occurs in stage s in $T_\mathfrak{R}$ and let m be the lowest stage at which every outgoing branch from $\mathfrak{R}^s$ has been closed. Then, the depth of $\mathfrak{R}^s$ is defined as $d = m - k$. With this it follows that if the reduction tree $T_\mathfrak{R}$ does not have a closed branch B where the topmost many-sided anti-sequent $\mathfrak{R}^B$ is an axiom of $\mathbf{MSR}_{\mathcal{L}_m}$, then $\mathfrak{R}$, the many-sided anti-sequent at the root of $T_\mathfrak{R}$, is irrefutable.

□

6 Conclusion

In this paper, we presented a systematic method for constructing many-sided anti-sequent calculi for finite-valued propositional logics, based on an approach as put forth by Zach (1993). With that, concrete sequent-type rejection calculi for particular logics from the literature can be directly obtained.

Another important type of calculi used in the context of many-valued logics are proof systems based on *hypersequents* (Avron, 1991a, 1996), which are tuples of two-sided sequents of the form $\Gamma_1 \vdash \Delta_1 \mid \cdots \mid \Gamma_n \vdash \Delta_n$, where "|" is interpreted disjunctively, i.e., such a hypersequent is true if at least one component sequent is true (under a respective semantics), and with suitable rules manipulating different components. It would be an interesting question for future work to build rejection systems for specific logics based on these kinds of sequents.

References

Avron A (1991a) Hypersequents, logical consequence and intermediate logics for concurrency. Annals of Mathematics and Artificial Intelligence 4:225–248

Avron A (1991b) Natural 3-valued logics - Characterization and proof theory. Journal of Symbolic Logic 56 (1):276–294

Avron A (1996) The method of hypersequents in the proof theory of propositional non-classical logics. In: Logic: From Foundations to Applications, Clarendon Press, New York, NY, USA, pp 1–32

Berger G, Tompits H (2014) On axiomatic rejection for the description logic $\mathcal{ALC}$. In: Declarative Programming and Knowledge Management–Declarative Programming Days (KDPD 2013), Revised Selected Papers, Springer, Lecture Notes in Computer Science, vol 8439, pp 65–82

Bonatti PA (1993) A Gentzen system for non-theorems. Tech. Rep. CD-TR 93/52, Christian Doppler Labor für Expertensysteme, Technische Universität Wien

Bryll G (1969) Kilka uzupełnień teorii zdań odrzuconych (Some supplements to the theory of rejected propositions). Zeszyty Naukowe Wyższej Szkoły Pedagogigicznej w Opolu, Seria B, Studia i Monografie 22:133–154

Bryll G, Maduch M (1969) Aksjomaty odrzucone dla wielowartościowych logik Łukasiewicza (Rejected axioms for Łukasiewicz's many-valued logics). Zeszyty Naukowe Wyższej Szkoły Pedagogicznej w Opolu, Matematyka VI:3–19

Dutkiewicz R (1989) The method of axiomatic rejection for the intuitionistic propositional logic. Studia Logica 48(4):449–459

Gentzen G (1935a) Untersuchungen über das logische Schließen I. Mathematische Zeitschrift 39(1):176–210

Gentzen G (1935b) Untersuchungen über das logische Schließen II. Mathematische Zeitschrift 39(1):405–431

Goranko V (1994) Refutation systems in modal logic. Studia Logica 53(2):299–324

Gödel K (1932) Zum intuitionistischen Aussagenkalkül. Anzeiger Akademie der Wissenschaften Wien, mathematisch-naturwissenschaftliche Klasse 32:65–66

Łukasiewicz J (1920) O logice trójwartościowej (On three-valued logic). Ruch filozoficzny 5:170–171

Łukasiewicz J (1921) Logika dwuwartościowa (Two-valued logic). Przegląd Filozoficzny 23:189–205

Łukasiewicz J (1939) O sylogistyce Arystotelesa (On Aristotle's syllogistic). Sprawozdania z Czynności i Posiedzeń Polskiej Akademii Umiejętności 44

Łukasiewicz J (1957) Aristotle's Syllogistic from the Standpoint of Modern Formal Logic, 2nd edn. Clarendon Press, Oxford

Oetsch J, Tompits H (2011) Gentzen-type refutation systems for three-valued logics with an application to disproving strong equivalence. In: Proceedings of the 11th International Conference on Logic Programming and Nonmonotonic Resoning (LP-NMR 2011), Springer, Lecture Notes in Computer Science, vol 6645, pp 254–259

Pinto L, Dyckhoff R (1995) Loop-free construction of counter-models for intuitionistic propositional logic. In: Symposia Gaussiana, Walter de Gruyter & Co., pp 225–232

Post E (1920) Introduction to a general theory of elementary propositions. Bulletin of the American Mathematical Society 26:437

Rosser JB, Turquette AR (1952) Many-valued logics. North-Holland, Amsterdam

Rousseau GS (1967) Sequents in many valued logic I. Fundamenta Mathematicae 60:23–33

Schütte K (1956) Ein System des verknüpfenden Schließens. Archiv für mathematische Logik und Grundlagenforschung 2(2-4):55–67

Skura T (1989) A complete syntactic characterization of the intuitionistic logic. Reports on Mathematical Logic 23

Skura T (1993) Some results concerning refutation procedures. Acta Universitatis Wratislaviensis, Logika 15

Skura T (1999) Aspects of refutation procedures in the intuitionistic logic and related modal systems. Acta Universitatis Wratislaviensis, Logika 20

Skura T (2011) Refutation systems in propositional logic. In: Handbook of Philosophical Logic, Second Edition, vol 16, Springer, pp 115–157

Skura T (2013) Refutation Methods in Modal Propositional Logic. Semper, Warszawa

Słupecki J (1948) Z badań nad sylogistyką Arystotelesa. Prace Wrocławskiego Towarzystwa Naukowego (B) 6, english translation: Słupecki (1951)

Słupecki J (1951) On Aristotelian syllogistic. Studia Philosophica, Posnaniae 4:275–300

Słupecki J (1959) Funkcja Łukasiewicza (Łukasiewicz's function). Zeszyty Naukowe Uniwersytetu Wrocławskiego, Seria A 3:33–40

Słupecki J, Bryll G, Wybraniec-Skardowska U (1971) Theory of rejected propositions I. Studia Logica 29(1):75–115

Słupecki J, Bryll G, Wybraniec-Skardowska U (1972) Theory of rejected propositions II. Studia Logica 30(1):97–139

Tiomkin ML (1988) Proving unprovability. In: Symposium on Logic in Computer Science, IEEE Computer Society, pp 22–26

Wybraniec-Skardowska U (1969) Teoria zdań odrzuconych (Theory of rejected sentences). Zeszyty Naukowe Wyższej Szkoły Pedagogigicznej w Opolu, Seria B, Studia i Monografie 22:5–131

Wybraniec-Skardowska U (2005) On the notion and function of the rejection of propositions. Acta Universitatis Wratislaviensis, Logika 23:179–202

Zach R (1993) Proof theory of finite-valued logics. Master's thesis, Technische Universität Wien, Institut für Computersprachen

www.ingramcontent.com/pod-product-compliance
Lightning Source LLC
Chambersburg PA
CBHW070657100426
42735CB00039B/2196